THE
FOURTH
DIMENSION

And How to Get There

THE
FOURTH
DIMENSION

And How to Get There

RUDY RUCKER

Foreword by Martin Gardner

Illustrations by David Povilaitis

RIDER

London Melbourne Sydney Auckland Johannesburg

Rider and Company

An imprint of the Hutchinson Publishing Group

17–21 Conway Street, London W1P 6JD

Hutchinson Publishing Group (Australia) Pty Ltd
16–22 Church Street, Hawthorn, Melbourne, Victoria 3122

Hutchinson Group (NZ) Ltd
32–34 View Road, PO Box 40-086, Glenfield, Auckland 10

Hutchinson Group (SA) Pty Ltd
PO Box 337, Bergvlei 2012, South Africa

First published 1985
© Rudy Rucker 1985

Printed and bound in Great Britain by Anchor Brendon Ltd,
Tiptree, Essex

ISBN 0 09 159930 X

For A Square, on his hundredth anniversary

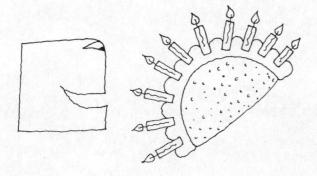

Contents

Foreword
by Martin Gardner

As WRITERS, mathematicians are notoriously inept. There are happy exceptions, of course, and at least one teacher of mathematics, Lewis Carroll, wrote immortal fantasy fiction. Eric Temple Bell not only wrote colorfully about mathematics for laymen, but under the pseudonym John Taine he churned out a raft of science fiction novels. In more recent times, several professional mathematicians have written science fiction with strong mathematical underpinnings. Now along comes Rudolf von Bitter Rucker — mathematician, novelist, cartoonist, rock music buff, and a thinker with the courage to explore dark and unfamiliar territory in what he likes to call the mindscape.

Rucker's major mathematical interests are transfinite sets (he has a doctorate in mathematical logic from Rutgers) and spaces of higher dimensions. After editing a selection of Charles Hinton's writings about the fourth dimension, and writing a popular account of four-space and relativity, Rucker's first big success came in 1982 with the publication of *Infinity and the Mind*. Science fiction fans had earlier known him for his wild, funny, sexually outrageous novels and stories in which higher spaces often play essential roles. His best-known novel, *White Light*, actually has the subtitle *What Is Cantor's Continuum Problem?* The subtitle is the exact title of a paper by the eminent logician Kurt Gödel, with whom Rucker had the privilege of many stimulating conversations.

The Fourth Dimension: Toward a Geometry of Higher Reality will be eagerly read by writers and readers of science fiction, but anyone with even a minimal interest in mathematics and fantasy will find the book as informative and mind-dazzling as Rucker's book on infinity. After exploring the whimsical underworld of Flatland, he plunges into spaces above three with a zest and energy that is breathtaking. The book is interspersed with clever problems for the mathematically competent, and finally, in a grand Carrollian climax, Rucker boldly invades the infinite dimensions of Hilbert space.

Science fiction? In part, yes, and Rucker pauses frequently to quote from some of his bizarre tales. But without the tool of infinite-dimensional spaces, modern physics would be almost impossible. Older books on quantum mechanics talk about how the measurement of particles and quantum systems "collapses the psi function." More recent books prefer the language of a complex Hilbert space, an approach adopted by John von Neumann in his classic work on quantum theory. When you measure a quantum system you are said to "rotate the state vector," an abstract line of precisely defined length and orientation that represents the system's state in a set of coordinate systems that constitute a Hilbert space. Are such spaces "real"? Or are they no more than convenient fictions used by physicists to simplify their calculations?

It is with such deep ontological questions that Rucker concerns himself throughout this book and especially in the final chapters. Here I am unable to go along with his "All is One" philosophy. He seems to have inherited a genetic fondness for the Absolute from his great-great-great-grandfather, the famous philosopher Hegel. Like William James, I do not know if ultimate reality is One or Many. Nor can I accept Rucker's views on synchronicity, taken over from Jung and Koestler, and his seeming belief that consulting the *I Ching* is more likely to produce meaningful "coincidences" than consulting say, Homer, the Bible, or the works of Isaac Asimov.

But no matter. We are each entitled to what James called our "over beliefs." Whether you agree or disagree with Rucker's Tao-tinged metaphysics, you will find his speculations twisting your mind toward fundamental questions that refuse to go away no matter how hard pragmatists and positivists try to banish them.

Preface

IN 1958, the Louisville Free Public Library had a single shelf of science fiction books. My friend Niles Schoening and I used to read and discuss them, wondering about time travel, wondering about the fourth dimension. This is where it all began.

When I went off to college in 1963, my father, Embry Rucker, gave me a copy of Edwin Abbott's *Flatland*. As an Episcopal priest, my father had already realized that the fourth dimension can serve as a symbol for higher spiritual realities.

In the ensuing years, I puzzled over the relationship between the fourth dimension as *higher reality*, and the fourth dimension as *time*. When I got my first teaching job, at SUNY Geneseo, I began working out these connections in my lectures for a course on higher geometry. These lectures were published by Dover Publications in 1977, under the title *Geometry, Relativity and the Fourth Dimension*.

In the seven years since then, I've learned a lot more about the fourth dimension. In writing *The Fourth Dimension: Toward a Geometry of Higher Reality*, I've tried to present a definitive and popular account of what the fourth dimension means, both physically and spiritually. I am grateful to Martin Gardner for lending me a number of rare books on the fourth dimension; to Thomas Banchoff for his help with the more technical research; to my editor, Gerard Van der Leun, for his encouragement; and to my illustrator, David Povilaitis, for his wit and weirdness.

Most of all, I would like to thank my family, my friends, my students, and my correspondents. Enjoy the book!

Part I

THE
FOURTH
DIMENSION

1

A New Direction

IS THIS ALL THERE IS? Struggle, loneliness, disease, and death . . . Is this all there is? Life can seem so chaotic, so dreary, so grindingly hard. Who among us has not dreamt of some higher reality, some transcendent level of meaning and peace?

There actually *is* such a higher reality . . . And it is not so very hard to reach. For many, the fourth dimension has served as a gateway into it. But what is the fourth dimension?

No one can point to the fourth dimension, yet it is all around us. Philosophers and mystics meditate upon it; physicists and mathematicians calculate with it. The fourth dimension is part and parcel of many respected scientific theories, yet it is also of great use in such disreputable fields as spiritualism and science fiction.

The fourth dimension is a direction different from all the directions in normal space. Some say that time is the fourth dimension . . . And this is, in a sense, true. Others say that the fourth dimension is a hyperspace direction quite different from time . . . This is also true.

There are, in fact, many higher dimensions. One of these higher dimensions is time, another higher dimension is the direction in which space is curved, and still another higher dimension may lead toward some utterly different universes existing parallel to our own.

At the deepest level, our world can be regarded as a pattern in infinite-dimensional space, a space in which we and our minds move like fish in water.

Fig. 1. A new answer for old questions.

Fig. 2. Where is it?

Ordinarily, of course, we say that we live in three-dimensional space. What precisely is meant by this? Why *three*? Watch the wheeling flight of swallows chasing gnats at dusk. Mathematically speaking, these lovely sweeping curves are of great complexity. But it is possible to break any such space curve down into three types of motion: east / west, north / south, and up / down. By combining the three mutually perpendicular types of motion, one can trace out any possible curve in our space. *No more than three* directions are needed, and *no less than three* directions will do — hence we call our space *three*-dimensional.

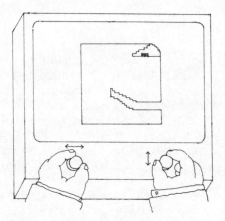

Fig. 3. Etch-A-Sketch drawing of A Square.

And when we shall see or feel ourselves in the world of four dimensions we shall see that the world of three dimensions does not really exist and has never existed; that it was the creation of our own fantasy, a phantom host, an optical illusion, a delusion — anything one pleases excepting only reality.

P. D. OUSPENSKY,
Tertium Organum, 1912

This fact is illustrated in two dimensions by a toy that was popular a few years ago, the Etch-A-Sketch. The underside of the Etch-A-Sketch's glass screen is covered by some silvery dust. Turning the knobs moves a stylus under the screen, and the stylus scrapes off dust, leaving dark trails. The left-hand knob moves the stylus in the left / right direction, and the right-hand knob moves the stylus in the up / down direction. If one twists the two knobs at the same time, one can draw any two-dimensional curve at all.

It is not too hard to imagine a three-dimensional Etch-A-Sketch that would, let us say, move a brightly flaring sparkler about in a dark room. As the image of a sparkler stays on the retina for a few seconds, one could thus have the experience of seeing three-dimensional curves generated by twiddling three different knobs: left / right, up / down, and back / forth.

Speaking of sparklers, there is a nice picture inside the cover of the Rolling Stones album *Black and Blue.* It is a time exposure of the five Stones waving sparklers. Bill Wyman traces a flat, tightening spiral. Ronnie Wood produces a messy figure eight. Charlie Watts slowly and patiently draws a big letter *O.* Keith starts out high and lets the sparkler fall in a tired zigzag. And Mick . . . Ah, Mick . . . Mick traces the only truly three-dimensional curve in the group: a complex rodeo pattern of swoops and loops. Waving sparklers in the dark is a good way to really savor our space's three-dimensionality.

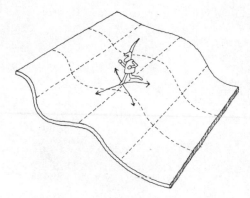

Fig. 4. Two degrees of freedom on a curved surface.

Another way of expressing all this is to say that motion in our space has *three degrees of freedom.* At any instant, a bird has three essentially different ways to alter its flight: speed up / slow down, wheel left / wheel right, climb / dive. Although we can wave our sparklers with just as much freedom, we cannot really move our bodies around in this way. Someone hiking in the hills moves up and down with the roll of the land . . . Yet in terms of control, he has only *two* degrees of freedom: forward / backward, and left / right. One can, of course, jump up and down a bit, but because of gravity, the effects of this are more or less negligible.

The point I am making here is that in terms of degrees of

Let us assume that the three dimensions of space are visualized in the customary fashion, and let us substitute a color for the fourth dimension. Every physical object is liable to changes in color as well as in position. An object might, for example, be capable of going through all shades from red through violet to blue. A physical interaction between any two bodies is possible only if they are close to each other in space as well as in color. Bodies of different colors would penetrate each other without interference . . . If we lock a number of flies into a red glass globe, they may yet escape: they may change their color to blue and are then able to penetrate the red globe.

HANS REICHENBACH,
The Philosophy of Space and Time, 1927

freedom, motion on the Earth's bumpy surface is basically two-dimensional. The surface itself is a curved three-dimensional object, granted. But any motion that is confined to this surface is essentially a two-dimensional motion. It could be that mankind's perennial dream of flight is a hunger for more dimensions, for more degrees of freedom. The average person only experiences three-dimensional body motion when he or she swims underwater.

Driving a car involves sacrificing yet another degree of freedom. One speeds up or slows down (possibly even reversing direction), but that's all. The road itself is a space curve in three-dimensional space, but motion that is confined to this particular curve is basically one-dimensional.

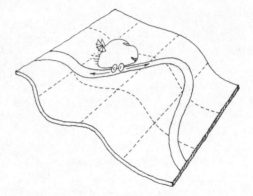

Fig. 5. One degree of freedom on a curved line.

As we will see later, the space we live in is also curved: curved like a hillside, twisted like a mountain road. But, in terms of *degrees of freedom*, it is clear that our space is three-dimensional. Another way of expressing this is to point out that we can give any location above the Earth's surface by using *three* numbers: longitude, latitude, and height above sea level. Again, if we're both in a city, I might typically tell you how to find me by giving you *three* bits of information. "Walk four blocks uptown, turn right and go two blocks crosstown, then go in the building there and ride the elevator to the twenty-fifth floor."

Now, if our space were four-dimensional, such instructions would usually need a fourth component. "Get out of the elevator and shift through six levels of reality." Just as there are many floors over a given location on a city's two-dimensional street grid, we can abstractly imagine there being many different "reality levels" available at each of our space locations. In a sense this is true . . . Even though we're in the same room, I might ask, "Where's your head at?" in an effort to get into closer contact.

Let us pursue this line of thought a bit further. Imagine that objects in space could exist at different reality levels, and to make it quite concrete, suppose that each level has its characteristic color — ranging from red through green to blue. Assume that objects interact only with objects of the same color. A person on the twenty-fifth floor will not stumble over someone on the second floor; we propose that a blue person can pass right through a green person.

In this example, *reality level* or *color* would constitute a fourth dimension. The three space dimensions plus the color dimension would make up a sort of four-dimensional space. A typical person would probably exist on several levels at once. Waving a four-dimensional sparkler here would involve having the light's color, or reality level, change in some complicated way. This would be one way of beginning to think about a four-dimensional space.

Another, somewhat similar, approach is to propose using *time* as a fourth dimension. If, after all, I really want to see you, it is not enough to tell you how many blocks and how many floors to travel. I need to tell you *how soon* to show up. Maybe I will not be at the rendezvous for another hour . . . And maybe I'll then stay there for only fifteen minutes. To really specify an event, it is not enough to give its longitude, latitude, and height above sea level. One must also state *when* it occurs. Just as a blue person can walk through a green person, a 2:00 A.M. person can walk through a 6:00 P.M. person. In terms of the waving sparklers, the dimension of time comes into play when one notes how *rapidly* the sparkler moves along each part of its path.

But somehow it misses the mark to represent the fourth dimension by reality level, by color, or by time. What is really needed here is the concept of a fourth *space* dimension. It is very hard to visualize such a dimension directly. Off and on for some fifteen years, I have tried to do so. In all

SOCRATES: And now, let me show in a figure how far our nature is enlightened or unenlightened: — Behold! human beings living in an underground den, which has a mouth open towards the light and reaching all along the den; here they have been from their childhood, and have their legs and necks chained so that they cannot move, and can only see before them, being prevented by the chains from turning round their heads. Above and behind them a fire is blazing at a distance, and between the fire and the prisoners there is a raised way; and you will see, if you look, a low wall built along the way, like the screen which marionette players have in front of them, over which they show the puppets.

GLAUCON: I see.

S: And do you see men passing along the wall carrying all sorts of vessels, and statues and figures of animals made of wood and stone and various materials, which appear over the wall? Some of them are talking, others silent.

G: You have shown me a strange image, and they are strange prisoners.

S: Like ourselves; and they see only their own shadows, or the shadows of one another, which the fire throws on the opposite wall of the cave?

G: True; how could they see anything but the shadows if they were never allowed to move their heads?

S: And of the objects which are being carried in like manner they would only see the shadows?

G: Yes.

this time I've enjoyed a grand total of perhaps fifteen minutes' worth of direct vision into four-dimensional space. Nevertheless, I feel that I understand the fourth dimension very well. How can this be? How can we talk productively about something that is almost impossible to visualize?

The key idea is to reason by analogy. The fourth dimension is to three-dimensional space as the third dimension is to two-dimensional space. 4-D : 3-D :: 3-D : 2-D. This particular analogy is one of the oldest head tricks known to man. Plato was the first to present it, in his famous allegory of the cave.

Fig. 6. Plato's Cave.

Here Plato asks us to imagine a race of men who are chained up in an underground den, chained in such a way that all they can ever look at is shadows on their cave's wall. Behind the men is a low ramp, and behind that a fire. Objects move back and forth on the ramp, and the fire casts shadows of these objects on the cave's wall. The prisoners think that these shadows are the only reality . . . They do not even realize that they have three-dimensional bodies. They talk to each other, but hearing the echoes bounce off the wall, they assume that they and their fellows are also shadows.

There are several interesting features in Plato's allegory. It is particularly striking that the prisoners actually think that they are their own shadows. This is interesting because it suggests the idea that a person is *really* some higher-dimensional soul that influences and watches this "shadow world" of three-dimensional objects.

To bring this particular idea home, let us update Plato's cave allegory a bit. Imagine a very large TV screen that displays full-color computer-generated images of people and objects moving about. Now imagine some people who have from birth been chained motionless in front of the giant tube. Electrodes run from their nervous systems to the image-generating computer, and for each person there is a particular TV personality he or she can control. These prisoners would mistake the flat phosphorescent TV screen for reality.

s: And if they were able to converse with one another, would they not suppose that they were naming what was actually before them?

G: Very true.

s: And suppose further that the prison had an echo which came from the other side, would they not be sure to fancy when one of the passers-by spoke that the voice which they heard came from the passing shadow?

G: No question.

s: To them, the truth would be literally nothing but the shadows of the images.

G: That is certain.

PLATO,
The Republic, circa 370 B.C.

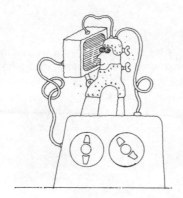

Fig. 7. Plato's Cave II.

So, one conclusion to draw from Plato's allegory is that we should not be too sure that our everyday view of the world is the most correct and most comprehensive view possible. Common sense can be misleading, and there may be a great deal more to reality than meets the eye.

PUZZLE 1.1

Stare out your window and imagine that the objects you see are actually two-dimensional shapes embedded in the window glass. The window glass is thus a sort of two-dimensional world. Under what condition can, say, two car shapes move through each other without collision!

An even more important aspect of Plato's allegory is that this allegory introduces the notion of a two-dimensional world. Insofar as the prisoners in the cave really think that they are shadows on the wall, they are viewing themselves as two-dimensional patterns. What would it be like to be a two-dimensional being? Would a two-dimensional being be able to imagine a third dimension?

Fig. 8. A Square.

In the next chapter we will talk about an imaginary two-dimensional world known as Flatland; and we will study the adventures of A Square, Flatland's most famous citizen. A Square's path to an understanding of the third dimension is, we will see, a guide for our own attempts to understand the fourth dimension.

2

Flatland

F*LATLAND,* first published in 1884, is the story of a square who takes a trip into higher dimensions. A century has passed and people are still talking about it. The author of *Flatland* was a Victorian schoolmaster named Edwin Abbott Abbott. Given the curious fact that his middle and last names were identical, it seems possible that Abbott might have been nicknamed *Abbott* Squared or *A* Squared. Thus, it may be that Abbott felt a considerable degree of identification with A Square, the hero of *Flatland.* After all, Abbott's life was, in some ways, as strictly regulated as the life of a two-dimensional Flatlander.

Edwin Abbott Abbott was born in London on December 20, 1838, the son of Edwin Abbott, head of the Philological School at Marylebone. Abbott attended the City of London School as a boy. He went on to Cambridge, was ordained a minister, was married, and at the age of twenty-seven returned to the City of London School as headmaster. He wrote a number of books on grammar and theology, books with titles like *How to Parse* and *Letters on Spiritual Christianity. Flatland* was his one venture into fantasy.

The book works on three levels. Most obviously, it is a satire on the staid and heartless society of the Victorians. "Irregulars" (cripples) are put to death, women have no rights at all, and when A Square tries to teach his fellows about the third dimension he is imprisoned. The second level of meaning in *Flatland* is scientific. By thinking about

Edwin Abbott Abbott (1838–1926)

I call our world Flatland, not because we call it so, but to make its nature clearer to you, my happy readers, who are privileged to live in Space.

Imagine a vast sheet of paper on which straight Lines, Triangles, Squares, Pentagons, Hexagons, and other figures, instead of remaining fixed in their places, move freely about, on or in the surface, but without the power of rising above or sinking below it, very much like shadows — and you will then have a pretty correct notion of my country and countrymen. Alas, a few years ago, I should have said "my universe" but now my mind has been opened to higher views of things.

EDWIN A. ABBOTT, *Flatland*, 1884

A Square's difficulties in understanding the third dimension, we become better able to deal with our own problems with the fourth dimension. Finally, at the deepest level, we can perhaps view *Flatland* as Abbott's circuitous way of trying to talk about some intense spiritual experiences. A Square's trip into higher dimensions is a perfect metaphor for the mystic's experience of higher reality.

Flatland is a plane inhabited by creatures that slide about. We might think of them as being like coins on a tabletop. Alternatively, we could think of them as colored patterns in a soap film, or ink spots in a sheet of paper.

In Flatland, the lower classes are triangles with only two sides equal. The upper classes are regular polygons, that is, figures with all sides equal. The more sides one has, the greater one's social standing. The highest caste of all consists of polygons with so many sides that they are indistinguishable from perfect circles.

As mentioned above, *Flatland* is more than just a book about dimensions. In somewhat the same manner as *Gulliver's Travels*, it satirizes the attitudes of the society in which its author lived. In our Western culture, women have probably never been at such a disadvantage as in the nineteenth century. Accordingly, the women of Flatland are not even skinny triangles: they are but lines, infinitely less respected than the priestly circles. Abbott, of course, realizes the injustice of this. When a sphere from "Space-land" visits Flatland, he has this to say: "It is not for me to classify human faculties according to merit. Yet many of the best and wisest in Spaceland think more of the affections than of the understanding, more of your despised Straight Lines than of your belauded Circles."

I take the plane upon which my shadow-man is, and move it through three dimensions. Thus the shadow-man perceives this third dimension. The man himself may be changed, and, at the end of his trip be pale and rumpled, when he had started the trip rosy and flat.

GUSTAV THEODOR FECHNER, "Why Space Has Four Dimensions," 1846

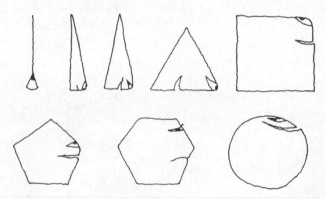

Fig. 9. Eight Flatlanders: Woman, Soldier, Workman, Merchant, Professional Man, Gentleman, Nobleman, High Priest.

An initial question about Flatland is the problem of how these lines and polygons can see anything at all. If you were to put a number of cardboard shapes on a tabletop and then lower your eye to the plane of the table, you would really see just a bunch of line segments. How can Flatlanders tell a triangle from a square? How do they build up the idea of a two-dimensional world from their one-dimensional retinal images?

Abbott reports that the space of Flatland is permeated with a thin haze. Because of this, the glowing sides of the polygons shade off rapidly into dimness. If you are looking at the corner of a triangle and the corner of a pentagon, you can tell them apart because the triangle's sides shade off more rapidly.

This may seem a bit artificial, but just stop to consider this: *our* retinal images of the world are two-dimensional patterns, yet we can distinguish a wide range of three-

This great man was used to say that, as we can conceive beings (like infinitely attenuated bookworms in an infinitely thin sheet of paper) which possess only the notion of space of two dimensions, so we may imagine beings capable of realising space of four or a greater number of dimensions.

SARTORIUS VON WALTERSHAUSEN,
Biography of Carl Friedrich Gauss, 1860

dimensional objects. If I look, for instance, at a sphere and a flat disk, I can tell them apart by their shading. Another important way in which we notice our world's three-dimensionality is by the fact that objects can move *behind* and *in front of* each other. If, looking out a restaurant window, I see a person walk in front of my car, I do not assume that he is somehow dematerializing my car. I recognize that there is a third dimension of space, and that in this dimension the sidewalk is closer than the street. Just as we can build up a mental image of our three-dimensional world, the Flatlanders have adequate images of their two-dimensional world.

Fig. 10. Depth is simpler than dematerialization.

A Square's dimensional adventures begin when he has a dream, a dream of Lineland:

> I saw before me a vast multitude of small Straight Lines (which I naturally assumed to be Women) interspersed with other Beings still smaller and of the nature of lustrous points — all moving to and fro in one and the same Straight Line, and, as nearly as I could judge, with the same velocity.
>
> A noise of confused multitudinous chirping or twittering issued from them at intervals as long as they were moving; but sometimes they ceased from motion, and then all was silence.
>
> Approaching one of the largest of what I thought to be Women, I accosted her, but received no answer. A second and a third appeal on my part were equally ineffectual. Losing patience at what appeared to me intolerable rudeness, I brought my mouth into a position full in front of her mouth so as

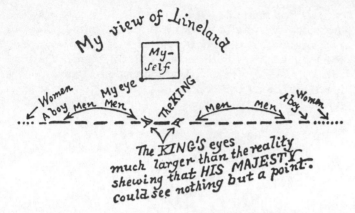

Fig. 11. Abbott's drawing of Lineland.

to intercept her motion, and loudly repeated my question, "Woman, what signifies this concourse, and this strange and confused chirping, and this monotonous motion to and fro in one and the same Straight Line?"

"I am no Woman," replied the small Line: "I am the Monarch of the world."

Although all the Linelanders can see of each other is a single point, they have a very good sense of hearing, and can estimate how far away each of their fellows is. The men have a voice at either end: bass on the left, tenor on the right. By noting the time lag between the two voices it is possible to tell how long a given male Linelander is. The poor women, of course, are just points!

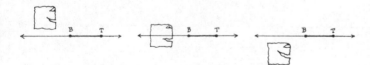

Fig. 12. A Square moves through Lineland.

A Square tries to tell the king about the second dimension. The king doesn't understand, and asks A Square to move in the direction of the mysterious second dimension. A Square complies, and moves right through the space of

I landed on all fours . . . There was a sort of floor about a yard below the plane of Flatland. When I stood up, it was as if I were standing waist-deep in an endless, shiny lake. My fall through the Flatlanders' space had smashed up one of their houses. Several of them were nosing at my waist, wondering what I was. To my surprise, I could feel their touch quite distinctly. They seemed to have a thickness of several millimeters . . .

I was standing in the middle of a "street," that is to say, in the middle of a clear path lined with Flatland houses on either side. The houses had the form of large squares and rectangles, three to five feet on a side. The Flatlanders themselves were as Abbott has described them: women are short Lines with a bright eye at one end, the soldiers are very sharp isosceles Triangles, and there are Squares, Pentagons and other Polygons as well. The adults are, on the average, about twelve inches across.

Lineland. (In figure 12 I have labeled the king's bass and tenor ends.) Naturally enough, the king simply perceives this "motion" as a segment that appears out of nowhere, stays for a minute, and then disappears all at once. The king denies the reality of the second dimension, A Square loses his temper, the dream ends.

Fig. 13. A Square and his wife in a locked room.

The next evening A Square and his wife are comfortably sealed up in the safety of their home, when suddenly a voice out of nowhere speaks to them. And then, a moment later, a circle appears in the confines of their tightly locked house. It is A Sphere, come to teach A Square about the third dimension.

Reasoning by analogy, you can see that a four-dimensional creature would be able to reach into any of our rooms

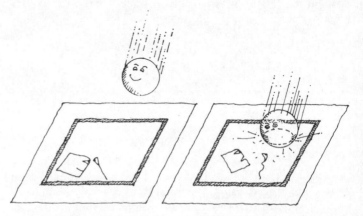

Fig. 14. A Circle appears in A Square's locked room.

or cubbyholes, no matter how well they are sealed up. A four-dimensional creature could empty out a safe without cracking it, for the safe has no walls against the fourth dimension. A four-dimensional surgeon could reach into your viscera without breaking your skin. A four-dimensional creature could drink up your Chivas Regal without ever opening the bottle!

The buildings that lined my street bore signs in the form of strings of colored dots along their outer walls. To my right was the house of a childless Hexagon and his wife. To my left was the home of an equilateral Triangle, proud father of three little Squares. The Triangle's door, a hinged line-segment, stood ajar. One of his children, who had been playing in the street, sped inside, frightened by my appearance. The plane of Flatland cut me at the waist and arms, giving me the appearance of a large blob flanked by two smaller blobs — a weird and uncanny spectacle, to be sure.

RUDY RUCKER,
"Message Found in a Copy of Flatland," 1983

Fig. 15. Liquor thief from the next dimension.

If only you had the muscles to twitch part of your arm up into the fourth dimension, you could reach in "around" the window at Tiffany's and take out the biggest diamond on display. This *would not* be done by somehow having your arm turn into gas or a ray of light. The heist would be done by having your arm move up through the fourth dimension. The diamond would be brought out by lifting it up into the fourth dimension to get "around" the sheet of glass.

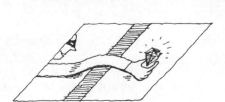

Fig. 16. The perfect crime.

Returning to A Square, there he is, locked in his house, talking to what seems to be a circle, another two-dimensional creature. The Sphere objects to this flat characterization of himself:

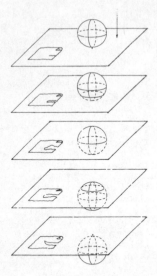

Fig. 17. A Sphere moves through Flatland.

"I am not a plane Figure, but a Solid. You call me a Circle; but in reality I am not a Circle, but an indefinite number of Circles, of size varying from a Point to a Circle of thirteen inches in diameter, one placed on the top of the other. When I cut through your plane as I am now doing, I make in your plane a section which you, very rightly, call a Circle. For even a Sphere — which is my proper name in my own country — if he manifest himself at all to an inhabitant of Flatland — must needs manifest himself as a Circle.

"Do you not remember — for I, who see all things, discerned last night the phantasmal vision of Lineland written upon your brain — do you not remember, I say, how when you entered the realm of Lineland, you were compelled to manifest yourself to the King, not as a Square, but as a Line, because that Linear Realm had not Dimensions enough to represent the whole of you, but only a slice or section of you? In precisely the same way, your country of Two Dimensions is not spacious enough to represent me, a being of Three, but can only exhibit a slice or section of me, which is what you call a Circle."

The Sphere proceeds to demonstrate the third dimension by moving through A Square's plane, just as A Square had moved through Lineland for the king. What A Square sees is a point that turns into a circle. The circle swells to a certain maximum size, then shrinks back to a point, which disappears. His great difficulty is in thinking of all these different circles as existing all together in the form of a sphere.

Pause for a moment and try to imagine four-dimensional space. It is right next to you, but in a direction you can't point to. No matter how well hidden you may be, a four-dimensional creature can see you perfectly well, inside and outside.

What would you see if, right this moment, a four-dimensional hypersphere were to pass through the space near your head? Reasoning strictly by analogy, you would expect to see first a point, then a small sphere, then a bigger sphere, then a small sphere, and then a final point that dis-

Fig. 18. A hypersphere moves through our space.

appears. Visually, it would be much the same as seeing a balloon that is first blown up and then deflated. Next time you have a balloon in hand, you might even try smoothly blowing it up and letting the air back out. That, basically, is what you would see if a hypersphere passed through the space of your room. A sphere is a three-dimensional stack of circles; a hypersphere is a four-dimensional stack of spheres.

But it is very hard to see how to stack things up in a new dimension. A Square, far from believing that he has seen the cross sections of a sphere, shrieks, "Monster, be thou juggler, enchanter, dream or devil, no more will I endure thy mockeries," and rams his hardest right angle against A Sphere's cross section.

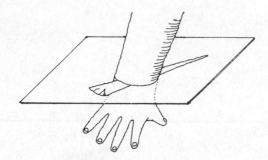

Fig. 19. Impaled by a policeman.

An interesting question comes up here. If A Square were actually to cut into A Sphere, would it matter to the sphere? Would it be possible for the Flatlanders to keep a three-dimensional being captive? To make this question quite concrete, imagine that you had somehow found the real Flatland. You stick your hand through it and an enraged isosceles triangle spears right through your wrist. Then what?

To answer the question, we have to decide exactly what Flatland is like. If the Flatlanders are truly two-dimensional, with no thickness at all, then they will be as immaterial as shadows or patches of light. In this case, having an isosceles intersect your wrist would not hurt, nor would it limit your freedom of movement. Indeed, it is debatable whether such an insubstantial Flatlander would actually be able to poke through your skin.

When residing and touring in the North of England several years ago, I talked and lectured several times on the fourth dimension. One day after having retired to bed, I lay fully awake thinking out some problems connected with this subject. I tried to visualize or think out the shape of a four-dimensional cube, which I imagined to be the simplest four-dimensional shape. To my great astonishment I saw before me first a four-dimensional globe and afterwards a four-dimensional cube, and learned only then from this object-lesson that the globe is the simplest body, and not the cube, as the third-dimensional analogy ought to have told me beforehand.

The remarkable thing was that the definite endeavor to see the one thing made me see the other. I saw the forms as before me in the air (though the room was dark), and behind the forms I saw clearly a rift in the curtains through which a glimmer of light filtered into the room . . .

I forgo the attempt to describe the four-dimensional cube as to its form. Mathematical description would be possible but would at the same time disintegrate the real impression in its totality. The fourth-dimensional globe can be better described. It was an ordinary three-dimensional globe, out of which, on each side, beginning at its vertical circumference bent, tapering horns proceeded, which, with a circular bend, united their points above the globe from which they started. The effect is best indicated

One problem with the idea of truly two-dimensional Flatlanders is that it is a little hard to see how they could have any solidity or reality. If they were indeed just like shaded regions of the plane, then there would really be nothing to prevent them from freely moving through each other. One way out of this difficulty might be to have the "atoms" of a two-dimensional Flatland consist of little wrinkles or bumps in the plane of the Flatland space. Thus A Square might be a sort of mesa in the rubbery sheet of Flatland's space. Here we could suppose that the space is infinitely thin, but it is easier to imagine that the "rubbery sheet of Flatland's space" does itself have a slight thickness.

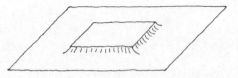

Fig. 20. A Square as a bump in the plane.

So now what if the Flatlanders *do* have a slight thickness? Abbott himself opts for this alternative in the preface to the second edition of *Flatland.* Here he describes how A Square comes to believe that insofar as Flatland really exists inside some higher space, it must be that the Flatlanders have height as well as length and breadth. Since each of them has the same height, there is no way for them to notice it. A Square reports an unintentionally funny conversation in which he discusses this point with the ruler of Flatland:

> I tried to prove to him that he was "high," as well as long and broad, although he did not know it. But what was his reply? "You say I am 'high'; measure my 'high-ness' and I will believe you." What could I do? How could I meet his challenge?

Even if all the Flatlanders were fully one inch high, they would not be able to notice this height, provided they still had no power of motion or variation in the third dimension. Of course, if there were Flatlanders of varying heights, then *some* kind of difference would be noticed, though the Flatlanders' description of the difference might be some vague, nongeometrical quality like charisma, force of personality, or "aura."

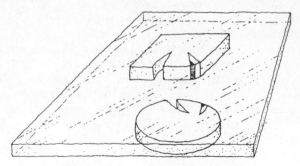

Fig. 21. A Square is high with the boss.

Now, if the Flatlanders did have an actual three-dimensional thickness, then having one of them cut into you would be like being cut with a knife blade. And if the creature was large enough, it might have enough mass to make your escape difficult.

In *Flatland*, as it turns out, Sphere becomes angry with Square's attempts to stab him, and finally grabs Square and lifts him up into space. A Square finds it an unsettling experience:

> An unspeakable horror seized me. There was a darkness; then a dizzy, sickening sensation of sight that was not like seeing; I saw a Line that was no Line; Space that was not Space: I was myself, and not myself. When I could find voice, I shrieked aloud in agony, "Either this is madness or it is Hell." "It is neither," calmly replied the voice of the Sphere, "it is Knowledge; it is Three Dimensions: open your eye once again and try to look steadily."

by circumscribing the numeral 8 by a circle. So three circles are formed, the lower one representing the initial globe, the upper one representing empty space, and the greater circle circumscribing the whole. If it now be understood that the upper circle does not exist and the lower (small) circle is identical with the outer (large) circle, the impression will have been conveyed, at least to some extent . . .

I have in a like manner had rare visions of the fifth- and sixth-dimensional figures . . . The fifth-dimensional vision is best described by saying that it looked like an Alpine relief map, with the singularity that all mountain peaks and the whole landscape represented in the map were *one* mountain, or again in other words as if all the mountains had one single base.

JOHAN VON MANEN,
Some Occult Experiences, 1913

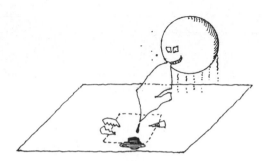

Fig. 22. What if the Sphere just got A Square's skin!

Among the countries of the world, Chile is remarkable for its great length and extreme narrowness. On the west coast of South America for 2700 miles it stretches ribbonlike from north to south, between the Andes Mountains and the Pacific Ocean. In some places it is less than 40 miles wide.

Let us suppose, for the sake of illustration, that the people of this land of Chile for some reason should be restricted from passing outside the boundaries of their country and in fact become incapable of having any communication with the outside world. Let us suppose also that the width of the country in an east and west direction gradually diminishes till it becomes so small as to be practically negligible. The Chilean world would thus become a very thin north and south vertical slice of space which would be two-dimensional for all practical purposes.

Correspondingly the inhabitants of the Chilean Thinland would become thin paper-like beings, capable of north and south and vertical motions, but not able to sidestep objects or to rotate the head sideways . . .

Here again, an interesting side issue comes up. Mightn't it be harmful for A Square to be lifted out of his space? We had better assume that the Square has some thin membranes sealing off his upper and lower faces against the third dimension, for otherwise when A Sphere pulls on a corner he might get only the skin! And rather than thinking of A Square as resting *on* his space, let's regard him as *in* a slightly thickened plane.

Fig. 23. A Square is really a thick part of a thick plane.

The best-known two-dimensional world other than Flatland is known as Astria. It is described in Charles H. Hinton's 1907 book, *An Episode of Flatland: Or How a Plane Folk Discovered the Third Dimension*. Here is the passage where Hinton explains his idea for the world:

> Placing some coins on the table one day, I amused myself by pushing them about, and it struck me that one might represent a planetary system of a certain sort by their means. This large one

A Square's gut busts him.

PUZZLE 2.1

It would seem that Flatlanders cannot have a complete digestive system in the form of a tube running the length of their bodies, for such a tube cuts them in half. Is there any way around this problem?

in the center represents the sun, and the others its planets journeying round it. And in this case considering the planets as inhabited worlds, confined in all their movements round the sun, to a slipping over the surface of the table, I saw that we must think of the beings that inhabit these worlds as standing out from the rims of them, not walking over the flat surface of them. Just as attraction in the case of our earth acts towards the center, and the center is inaccessible by reason of the solidity on which we stand, so the inhabitants of my coin worlds would have an attraction proceeding out in every direction along the surface of the table from the center of the coin, and "up" would be to them out from the center beyond the rim, while "down" would be towards the center inwards from the rim. And beings thus situated would be rightly described as standing on the rim. [Most of Hinton's essays can be found in *The Selected Writings of C. H. Hinton*, a 1980 Dover anthology.]

A drawback with a world design like this is that the polygons have so much trouble moving past each other, building houses, and so on. Many of these difficulties are resolved in

[T]he two eyes of a Chilean Thinlander would take up positions on the front edge of the face, as for instance one above the other on the forehead, or in order to get a longer base triangulation in distance vision, one on the forehead and the other on the tip of the chin. In order to look backward, that is, behind one, since there could be no turning around of the body or head sidewise as such motion would involve a use of the third dimension, a long flexible neck would have to be developed so that the head could be tilted backwards into an upside-down position.

FLETCHER DURRELL,
Mathematical Adventures, 1938

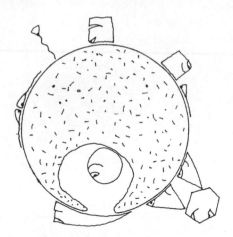

Fig. 24. Astrians on a small planet.

PUZZLE 2.2

Abbott's Flatland is not really a very close analogue of our world. For although our space is 3-D, we cannot move around freely in 3-D space. Instead we must walk around on the surface of a sphere. What would be an analogous design for a 2-D world?

A. K. Dewdney's 1984 book *The Planiverse*. As is *The Fourth Dimension*, Dewdney's book is in some measure a celebration of *Flatland*'s centennial.

One final plane-universe book worth mentioning is the 1965 book *Sphereland*, by the Dutch mathematician Dionys Burger. Burger describes a world that is something of a compromise between Flatland, with its great freedom of movement, and Astria, with its close resemblance to Earth. Burger's notion is that, as in Astria, the 2-D creatures live near the surface of a disk planet. But he proposes that they be very light, and thus able to live in their planet's atmosphere. It is as if people were able to live in clouds, clouds floating above tropical vegetation that is, in turn, floating on a sea that surrounds the planet's dense core.

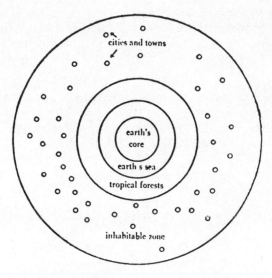

Fig. 25. Cities in the sky. (Drawing from Dionys Burger's Sphereland.*)*

3

Pictures of the Gone World

IMAGINE that you have been lifted up into hyperspace. What would our world look like from this vantage point? To begin with, note that our 3-D space would cut 4-D hyperspace into two regions, just as a 0-D point cuts a 1-D line in two, a 1-D line cuts a 2-D plane in two, and a 2-D plane cuts a 3-D space in two. (By the way, we speak of a point as *zero*-dimensional, 0-D, because someone whose entire space is limited to one point has *no* degrees of freedom in his or her motions.)

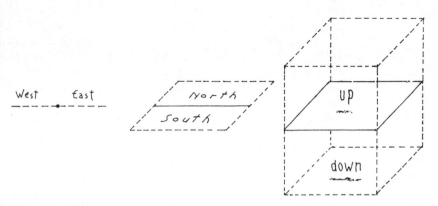

Fig. 26. An n-*dimensional space cuts an* (n + 1)-*dimensional space in half.*

Let us take a certain plane — for instance, that which separates the surface of Lake Ladoga which surrounds us, from the atmosphere above it, in this quiet autumn evening. Let us suppose that this plane is a separate world of two dimensions, peopled with its own beings, which can move only on this plane . . .

Let us suppose that, having escaped from behind our Schlusselburg bastions, you went for a bathe in the lake.

As beings of three dimensions you also have two dimensions which lie on the surface of the water. You will occupy a definite place in the world of shadow beings. All the parts of your body above and below the level of the water will be imperceptible to them, and they will be aware of nothing but your contour, which is outlined by the surface of the lake. Your contour must appear to them as an object of their own world, only very astonishing and miraculous. The first miracle from their point of view will be your sudden appearance in their midst. It can be said with full conviction that the effect you would create would be in no way inferior to the unexpected appearance among ourselves of some ghost from the unknown world. The second miracle would be the surprising changeability of your external form. When you are immersed up to your waist your form will be for them almost elliptical, because only the line on the surface surrounding your waist and impenetrable for them will be perceptible to them. When you begin to swim you will assume in their eyes the outline of a man.

What shall we call the two regions of hyperspace determined by our space? Charles H. Hinton has suggested the words *ana* and *kata*, to be used more or less like the words *up* and *down*. Just to have something to think about, we might think of heaven as lying *ana* above our space, and hell as lying *kata* below. A 4-D angel expelled from heaven would tumble through our space like a man falling through Flatland: an exciting moment of grotesque, incomprehensible cross sections splitting and merging and falling about!

Fig. 27. Man falling through Flatland.

Just as man's Flatland cross section could be a number of irregular shapes with skin boundaries, a hyperbeing's cross section in our space might be a bunch of bobbing globs of skin-covered flesh. Some of the blobs might have things like teeth or claws! Being "picked up" by a hyperbeing would probably involve a bunch of globs closing in on you like cross sections of a hand's fingers.

Once you are out in hyperspace, you can get some very strange perspectives on those you left behind. Consider how Flatland looks to us: we can see all four sides of A Square, and we can see every detail of innards. By the same token, a 4-D creature should be able to look down at me and, at one glance, see every square inch of my skin, the inside and outside of my stomach, the convolutions of my brain, and so on.

But how, you may ask, could a 4-D person "see" all sides

Fig. 28. Woman menaced by a creature from the fourth dimension.

When you wade into a shallow place so that the surface on which they live will encircle your legs, you will appear to them transformed into two ring-shaped beings. If, desirous of keeping you in one place, they surround you on all sides, you can step over them and find yourself free from them in a way quite inconceivable to them. In their eyes you would be an all-powerful being — an inhabitant of a higher world, similar to those supernatural beings about whom theologians and metaphysicians tell us.

N. A. MOROSOFF,
"Letter to My Fellow-Prisoners in the Fortress of Schlusselburg," 1891

of a 3-D object at once? A human being's retina is a two-dimensional disk of nerve endings. By analogy, we would expect a 4-D creature's retina to be a three-dimensional sphere of nerve endings. My seeing A Square consists of the excitation of a square-shaped pattern of nerve endings in my retina. A 4-D creature's "seeing" me would consist of the excitation of a person-shaped pattern of nerve endings in the little ball of his retina. Each point in A Square's body sends a light ray up to a single point in my retina. Each point in my body sends a light ray *ana* to a single point in the 4-D creature's retina.

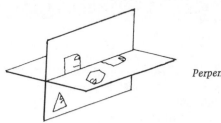

Perpendicular worlds.

PUZZLE 3.1

In four dimensions it is possible to have two 3-D spaces "perpendicular" to each other. Two such spaces would have only a plane in common. Suppose now that there is a 3-D space perpendicular to ours, a space with people moving around in it. Use a Flatland analogy to figure out how these people would appear to us.

I felt myself rising through space. It was even as the Sphere had said. The further we receded from the object we beheld, the larger became the field of vision. My native city, with the interior of every house and every creature therein, lay open to my view in miniature. We mounted higher, and lo, the secrets of the earth, the depths of mines and inmost caverns of the hills, were bared before me.

EDWIN A. ABBOTT, *Flatland*, 1884

Fig. 29. A 2-D retina with an image of A Square, and a 3-D retina with an image of a person.

It is a curious feature of 4-D space that we can connect two points in the interiors of two solid 3-D objects without piercing these objects' surfaces. The trick is to use *ana* / *kata* motions to get in and out of the solid 3-D objects. If you're inside a cubical room and move *ana* out of it, it's as if you had suddenly dematerialized. You don't go through the walls or the floor or the ceiling, you move over in the *ana* direction to a part of 4-D space where the room doesn't exist at all.

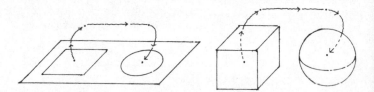

Fig. 30. Connecting interior points without cutting through the boundary.

So the reason that a 4-D creature can see all of me inside and out is that such a creature's "retina" is able to form a completely detailed model of my body. But this is not really

---

PUZZLE 3.2

If we assume that A Square's eye stays the same when he is lifted into 3-D space, he will not really be able to see whole 2-D objects as we do. What will he see? How might he build up a mental image of the full 2-D Flatland?

so puzzling or occult a phenomenon. The human brain is able to mimic such behavior . . . For do you not have a detailed 3-D mental image of your right hand? When you think of your hand you do not necessarily think of just the front or just the back. It is really possible to have the idea of a 3-D object seen from no particular direction — or from all directions at once.

We can form especially good 3-D images of transparent objects such as paperweights, wine bottles, or glasses of water. Here, unlike with the hand, there is no difficulty in imagining the inside parts. Holding 3-D images in your mind is definitely something worth doing. Try, for instance, to think about your house — all of it, seen from no particular vantage point. Here you are getting close to a higher-dimensional experience.

So a 4-D view of our 3-D world is not totally inconceivable. But what would it be like to look at 4-D objects? We saw with the hypersphere in chapter 2 that it is possible to get images of various 3-D cross sections of such a hyperobject, but how are we to combine these sections into a 4-D whole?

Some people might say at the outset that it is hopeless to try to think of four-dimensional objects. For how can our 3-D brains ever hold images of 4-D objects? This argument has some force, but it is not really conclusive. Drawings use 2-D arrangements of lines to represent 3-D objects. Why shouldn't we be able to build up 3-D arrangements of neurons that represent 4-D objects? More fancifully, perhaps our minds are not just 3-D patterns: maybe our brains have a slight 4-D hyperthickness; or maybe our minds extend out of our brains and into hyperspace!

I'll spend the rest of this chapter discussing two of the simpler 4-D shapes: the hypersphere and the hypercube. We'll start with the hypersphere, though if you hate math you might want to skip up to where I talk about the hypercube.

A sphere of any dimensionality is specified by giving its center and its radius. In any space, the sphere with center point O and radius r is the set of all points P whose distance from O is r. In 2-D space this definition leads to a circle of radius r, in 3-D space it gives a traditional sphere, and in 4-D space it gives a hypersphere.

Pick a point O in space near you and try to imagine a

The sphere of my vision now began to widen. Next I could distinctly perceive the walls of the house. At first they seemed very dark and opaque; but soon became brighter, and then *transparent:* and presently I could see the walls of the adjoining dwelling. These also immediately became light, and vanished, — melting like clouds before my advancing vision. I could now see the objects, the furniture, and persons, in the adjoining house as easily as those in the room where I was situated . . . But my perception still flowed on! The broad surface of the earth, for many hundred miles, before the sweep of my vision — describing nearly a semicircle — became transparent as the purest water; and I saw the brains, the viscera, and the complete anatomy of animals that were at that moment sleeping or prowling about in the forests of the Eastern Hemisphere, hundreds and even thousands of miles from the room in which I was making these observations.

ANDREW JACKSON DAVIS,
The Magic Staff, 1876

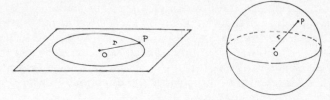

Fig. 31. A circle is a 2-D sphere.

hypersphere centered on this point, a hypersphere with a five-foot radius. What points P lie on the hypersphere? First of all, there are the points in your space that are five feet from 0. But moving in our space is not the only way to move away from 0. What if one combines motion away from 0, as before, with a motion *ana* out of our space? We might, for instance, move four feet away from 0 in our space, turn at right angles, and then move *ana* three feet into hyperspace. (Those who remember the Pythagorean theorem, or the "distance formula" of analytic geometry, can check that this is true because $4^2 + 3^2 = 5^2$.)

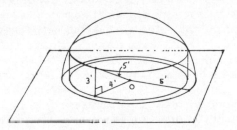

Fig. 32. 2-D plus "up" is like 3-D plus "ana."

What is interesting to notice here is that no matter which direction of our space the original four-foot displacement from 0 is taken in, the additional three-foot *ana* motion gives a point exactly five feet away from 0. So if we take all the points on a four-foot sphere around 0 and then move *ana* three feet, we will get a displaced sphere of points all belonging to the five-foot hypersphere around 0.

Now we can see why the full hypersphere consists of a series of spheres, spheres that grow smaller as one moves *ana* or *kata* from the space where the center lies. Taken together, this family of spheres makes up a three-dimensional "hypersurface," analogous to the two-dimensional surface of a sphere. The hypersurface of a hypersphere is a curved 3-D space located in 4-D space.

This is an important concept because many scientists believe that the space of our universe is in fact the hypersurface of a very large hypersphere. Let's try to understand it a little better.

First of all, shouldn't the hypersurface of a hypersphere be *four*-dimensional, not *three*? Not really. Consider the surface of an ordinary 3-D sphere such as the planet Earth. Although the surface is certainly curved in three dimensions, someone limited to the surface has only *two* degrees of freedom in moving: east/west or north/south. A Flatlander sliding around on a 3-D sphere's surface still feels himself to be in a 2-D space. It's just that this space somehow curves back on itself.

Now think of a little hyperfly who can move in hyperspace, but who has to stay exactly five feet from the point *0*. If the fly starts out five feet away from *0* in our space then it has basically three kinds of motion open to it: east/west or north/south (around a five-foot sphere centered on *0* in our space) or an *ana/kata* motion (combined with a motion toward *0* to keep the distance at five feet).

We'll come back to the hypersphere later, but now it's time to look at the *hypercube*.

The hypercube, also known as the *tesseract*, is probably the best-known 4-D geometrical pattern. It arises in the following way:

PUZZLE 3.3

See if you can complete this table:

	Corners	Edges	Faces	Solids
Point	1	0	0	0
Segment	2	1	0	0
Square	4	4	1	0
Cube				
Hypercube				
Hyperhypercube				

There wasn't much to the machine. All great things are simple, I suppose. There were three trussed beams of aluminum at right angles to each other, each with a cylinder and plunger, and, from them, toggles coming together at a point where there was a sort of "universal joint" topped by a mat of thick rubber. That was all . . .

So Banza stepped on the rubber mat and Bookstrom instructed him.

"Move this switch one button at a time. That will always raise you a notch. Look around each time until you get it just right."

With the first click Banza disappeared, just as people vanish suddenly in the movies. Cladgett groaned and squirmed and then was quiet. With another click Banza reappeared, and in his hand was a pair of old-fashioned pince-nez spectacles, moist and covered with

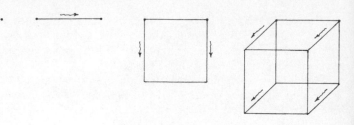

Fig. 33. From point to cube.

Start with a point and move it one unit to the right. This produces a one-dimensional line segment. Now move the line segment one unit down the page, producing a two-dimensional square. If we move the square one unit out of the page we get a three-dimensional cube.

Now, we can't really fit a three-dimensional object into the two-dimensional confines of this page. The standard convention, which we have used above, is to represent the third dimension as a direction diagonal to the first two. What if we were to use the *other* diagonal direction as the fourth dimension? If we move our image of the cube one unit in this "fourth dimension," we get a picture of a four-dimensional hypercube.

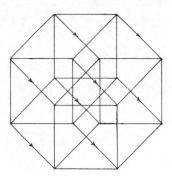

Fig. 34. The hypercube.

This figure is fun to look at . . . It has a certain mandala-like quality. If you are interested in drawing your own, note that the figure can be produced by constructing a square on each of the inner edges of a regular octagon. A regular octagon can be obtained by tearing down a STOP sign or, preferably, by dividing a circle into eight equal slices.

The hypercube arises as the "trail" of a cube moving in four-dimensional space. A cube arises as the "trail" of a square moving in three-dimensional space. Any given cube can be generated in three different ways, depending which of three possible pairs of opposite squares are thought of as being the "start" and "finish" positions. The hypercube includes four pairs of cubes. Can you see them all?

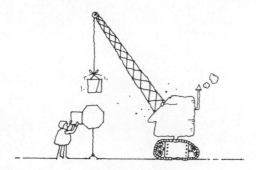

Fig. 35. *Squares on an octagon.*

A different way of drawing a hypercube is based on the idea that if you hold the wire skeleton of a cube close to your face, it will look like a small square inside a big square. Analogously, we can represent a hypercube by drawing a small cube inside a large cube. The idea is that the small cube is "farther away" in the direction of the fourth dimension. Surprisingly, such a hypercube shape can be naturally produced by dipping a cubical framework into soap solution, and then adding a bubble at the center.

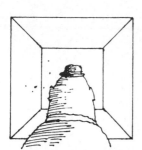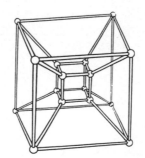

Fig. 36. *A cube as a square in a square, and a hypercube as a cube in a cube. (Hypercube drawing from D. Hilbert and H. Cohn-Vossen, Geometry and the Imagination.)*

a grayish film. He held them toward Cladgett, who grabbed them and mumbled something.

"Can you imagine," breathed Banza, "standing in the center of a sphere and seeing all the abdominal organs around you at once? Something like that, it seemed, not exactly either. There above my head were the coils of the small intestine. To the right was the cecum with the spectacles beside it, to my left the sigmoid and the muscles attached to the ilium, and beneath my feet the peritoneum of the anterior abdominal wall. But I was terribly dizzy for some reason; I could not stand it very long, much as I should have liked to remain inside of him for a while —"

MILES J. BREUER,
"The Appendix and the Spectacles," 1928

"All right. Now listen — a tesseract has eight cubical sides, *all on the outside.* Now watch me. I'm going to open up this tesseract like you can open up a cubical pasteboard box, until it's flat. That way you'll be able to see all eight of the cubes." Working very rapidly he constructed four cubes, piling one on top of the other in an unsteady tower. He then built out four more cubes from the four posed faces of the second cube in the pile. The structure swayed a little under the loose coupling of the clay pellets, but it stood, eight cubes in an inverted cross, a double cross, as the four additional cubes stuck out in four directions. "Do you see it now? It rests on the ground floor room, the next six cubes are the living rooms, and there is your study, up at the top."

Bailey regarded it with more approval than he had the other figures. "At least I can understand it. You say that is a tesseract, too?"

"That is a tesseract unfolded in three dimensions. To put it back together you tuck the top cube onto the bottom cube, fold those side cubes in till they meet the top cube, and there you are. You do all this folding through a fourth dimension of course; you don't distort any of the cubes, or fold them into each other."

Bailey studied the wobbly framework further. "Look here," he said at last, "why don't you forget about folding this thing up through a fourth dimension — you can't anyway — and build a house like this?"

ROBERT A. HEINLEIN,
"And He Built a Crooked House," 1940

Yet another way of representing a hypercube is by *unfolding* one. As before, we reason by analogy. If you cut some of the edges of a paper cube, you can flatten the cube down into a connected two-dimensional pattern of six squares. This can be done in eleven essentially different ways.

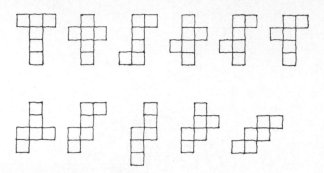

Fig. 37. The eleven ways to unfold a cube.

If a hypercube is cut in the proper way, then it can be unfolded and "flattened down" into a connected three-dimensional pattern of eight cubes. One unfolding of the hypercube produces a sort of three-dimensional cross. Salvador Dali actually uses this unfolded hypercube as a crucifix in his 1954 painting *Christus Hypercubus*. In his classic story "And He Built a Crooked House," Robert Heinlein describes a house built in this pattern. The twist in Heinlein's story occurs when there is an earthquake and the house folds up into a hypercube!

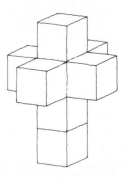

Fig. 38. One way to unfold a hypercube.

As a last approach to the hypercube, let's imagine a hollow stone hypercube, analogous to a hollow stone prison cell. What would we see if this hypercubical prison was pushed through our space?

Well, what would A Square see if we pushed a hollow stone cube through the plane of Flatland? At the beginning the whole solid stone floor would intersect his space, later there would be hollow stone squares made up of cross sections of the walls, and finally there would be the solid stone cross section of the ceiling. He would see a solid square, followed by hollow squares, followed by a solid square. By the same token, if we were to push a Flatland prison cell — a hollow stone square — through Lineland, the Linelanders would see a solid segment, hollow segments, and then a solid segment.

Reasoning by analogy, it is not hard to see that if a hollow stone hypercube were to pass through our space, we would see first a solid cube of stone, then a series of hollow stone

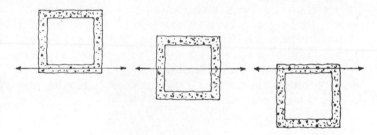

Fig. 39. A 2-D prison cell passes through Lineland.

PUZZLE 3.4

Figure 38 shows an unfolded hypercube. Try to figure out which sides should get glued to which if the thing is to be folded back up into a four-dimensional hypercube. Specifically, which faces get glued to the faces of the bottom cube?

PUZZLE 3.5

The volume of a cube S feet on a side is given by the formula S^3. What do you think would be the formula for the hypervolume of a hypercube S feet on a side? Specifically, what would be the hypervolume of a 2-by-2-by-2-by-2 hypercube?

cubes, and finally a last solid stone cube. The eight solid "boundary cubes" of the hypercube would arise as the two solid cubes seen at beginning and end, and as the six "trails" of the floors, walls, and ceilings of the hollow cubes.

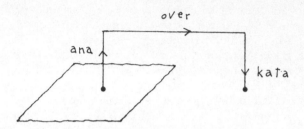

Fig. 40. Escaping from a sealed room.

The hollow stone hypercube could serve as a prison cell in which a hyperbeing, say an angel, could be incarcerated. Think of the room you are in as having a solid stone floor, walls, and ceiling. Even an angel can't get through that solid stone. Normally, however, an angel could escape the room by moving *ana*, over, and *kata*. But if your room is only a cross section of a stone hypercube, then when the angel moves *ana* he finds himself not in an empty space but still inside a stone-walled room. If the angel keeps moving *ana*

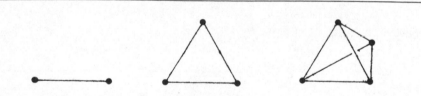

From segment to tetrahedron.

PUZZLE 3.6

The two points at either end of a line segment have the pleasant property of being equally far way from each other. If we move into 2-D space, we can find a third point so that now all three points are equally far away from each other. The three points, of course, are the vertices of an equilateral triangle. In 3-D space we can go out of the triangle's plane and get a fourth point so that now all four points are the same distance from each other. These four points make up the corners of a triangular pyramid, also known as a tetrahedron. What kind of 4-D figure do you get if you continue this procedure through one more step?

Fig. 41. No escape.

he will run into one of the boundary cubes: a whole roomful of solid stone, impossible to move beyond.

We have now discussed a variety of four-dimensional phenomena, but many readers may still feel that something is missing. One doesn't want only to *reason* about the fourth dimension, one wants somehow to *see* it. Most of what we have learned about the fourth dimension so far has been by using Lineland and Flatland analogies. Although the analogies are very instructive, it is not unusual to have the feelings P. D. Ouspensky describes in *Tertium Organum*.

> The method of analogies is, generally speaking, a rather tormenting thing. With it, you walk in a vicious circle. It helps you to elucidate certain things, and the relations of certain things, but in substance it never gives a direct answer to anything. After many and long attempts to analyze higher dimensions by the aid of the method of analogies, you feel the uselessness of all your efforts; you feel that you are walking alongside of a wall. Thereupon you begin to experience simply a hatred and aversion for analogies, and you find it necessary to search for the direct way which leads where you need to go.

In the next chapter we will look at a direct, though somewhat dangerous, path into the fourth dimension.

4

Through the Looking Glass

THE FOURTH DIMENSION is essentially a modern idea, dating back not much further than the midnineteenth century. Although the vague notion of a higher world has always been with us, the scientific concept of a geometrical fourth dimension was late to develop.

The first philosopher to seriously entertain the idea of higher-dimensional spaces was the great Immanuel Kant (1724–1804). In one of his earliest essays he writes longingly of such spaces: "A science of all these possible kinds of space would undoubtedly be the highest enterprise which a finite understanding could undertake in the field of geometry . . . If it is possible that there could be regions with other dimensions, it is very likely that God has somewhere brought them into being."

Later in his career, Kant proposed a famous puzzle that is related to the idea of the fourth dimension: If all of space were empty except for a single human hand, would it make sense to say that the hand was specifically a *right* hand? As it turns out, the answer is no. The concepts of "left" and "right" are meaningless in an empty space.

To begin to understand why, imagine a big Plexiglas sign advertising a palmistry shop, a shop run by the famous fortuneteller Mom Oxo. The outline and wrinkles of a palm are drawn on the transparent Plexiglas. Now, if you look at the sign from one side you see a right palm; if you look from the

Fig. 42. Kant's hand for Mom Oxo's Palmistry.

What can be more similar in every respect and in every part more alike to my hand and to my ear than their images in the mirror? And yet I cannot put such a hand as is seen in the glass in the place of its original; for this is a right hand, that in the glass is a left one, and the image or reflection of the right ear is a left one, which never can take the place of the other. There are in this case no internal differences which our understanding could determine by thinking alone. Yet the differences are internal as the senses teach, for, notwithstanding their complete equality and similarity, the left hand cannot be enclosed in the same bounds as the right one (they are not congruent); the glove of one hand cannot be used for the other. What is the solution?

IMMANUEL KANT,
Prolegomena to Any Future Metaphysics, 1783

other you see a left palm. Once you realize that it is possible to look from outside the two-dimensional plane of the sign, you see that it would make no sense to say the palm is really a right palm.

The same is true of three-dimensional space. Depending which four-dimensional "side" you look at space from, a hand can appear to be a left or a right. Another way of putting this is that a left hand can be converted into a right hand by lifting it into four-dimensional space and "turning it over."

Let's dramatize this in terms of A Square. At the end of *Flatland*, Square's trip into the third dimension ends with a big argument between him and A Sphere. Reasoning by analogy, A Square concludes that there must be a four-dimensional world beyond the sphere's 3-D space, and asks to be taken there. But here, as sometimes happens, the student is seeing much further than the teacher has ever dared. Sphere is annoyed and finally angered by Square's insistence that there must be a four-dimensional "Thoughtland" beyond Lineland, Flatland, and Spaceland. Yet A Square prattles on, dreaming of higher and higher dimensions. Sphere loses all patience and the trip is over:

> My words were cut short by a crash outside, and a simultaneous crash inside me, which impelled me through space with a velocity that precluded speech. Down! down! down! I was rapidly descending; and I knew that return to Flatland was my doom. One glimpse, one last and never-to-be-forgotten glimpse I had of that dull level wilderness — which was now to become my Universe again — spread out before my eye. Then a darkness. Then a final, all-consummating thunder-peal; and, when I came to myself, I was once more a common creeping Square, in my Study at home, listening to the Peace-Cry of my approaching Wife.

Now of course A Square wants to tell everyone about his revelations but — surprise, surprise — in Flatland it's illegal to talk about higher dimensions. Square tries to hold himself back, but finally at a meeting of the Local Speculative Society,

> I so far forgot myself as to give an exact account of the whole of my voyage with the sphere into Space . . . At first, indeed, I pretended that I was describing the imaginary experiences of a fictitious person but my enthusiasm soon forced me to throw off all disguise, and finally, in a fervent peroration, I exhorted all my hearers to divest themselves of prejudice and to become believers in the Third Dimension.
>
> Need I say that I was at once arrested and taken before the Council?

A Square is convicted and sentenced to life imprisonment. As time goes by he finds it harder and harder to think about the third dimension. Edwin Abbott's *Flatland* ends on this downbeat note, with the increasingly depressed Square having served out seven years of his sentence.

Nineteen eighty-four is the hundredth anniversary of *Flatland*'s publication. I am delighted to report that A Square is alive and flourishing. Not only is he alive, but he has consented to communicate an account of his further adventures. I will be quoting from my notes on *The Further Adventures of A Square* throughout the rest of this book. I will begin by quoting the first two paragraphs here, and will then proceed to some excerpts dealing with problems of left and right.

A full century has elapsed since last I communicated with your happy Spaceland race. My health, once failing, is again robust. Far from languishing in prison, I am now a respected Professor of Theology. The new Chief Circle encourages the Masses to worship Higher Space Beings as Angels and Gods. And my assistants are ever pressing forward the mathematical theory of many Dimensions.

My special Revelation, once so arcane, now suffers, if anything, from too great a currency among the Vulgar, and from too erudite an analysis among the Clever. Let us never lose sight of the fact that Higher Space is a royal road to that which is beyond all imagining. He that hath ears, let him hear.

At first I took the intruder for a fellow Square. But, feeling his Perimeter, I discovered something quite Irregular: he had no eye. I thought to console him, and the following conversation ensued.

I. Have they then imprisoned you for Blindness? And why in my cell?

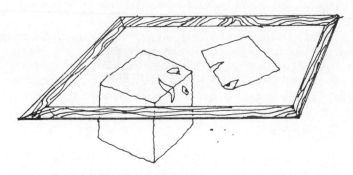

Fig. 43. A Cube visits A Square.

STRANGER. Lock me up? Not on your life, Square. I'm a Cube from Spaceland. Pleased to meet you.

I. Oh blessed Providence, can it be? But what has become of my old Mentor, the Sphere?

CUBE. The Sphere doesn't care, Square. If he did he'd have rescued you long ago. How much time have you served?

I. You inquire regarding the length of my imprisonment thus far? Seventy years, my Lord. I used, it is true, to wonder why A

Sphere did not lift me up out of my cell. But if I escaped, the Council would only reimprison me, or worse.

CUBE. Don't worry. I've been thinking what I could do for you. Cube and Square, they go together, right? My idea is that I do something to you that *proves* there's a third dimension.

I. Your meaning is yet veiled, my Lord.

CUBE. Well, check *this* out, cousin.

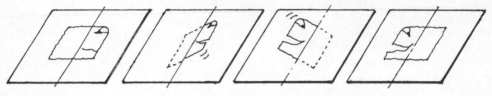

Fig. 44. A Square turns over.

The Cube lunged at me and seized my Corner in his mouth. I felt a strange whirling about my center and then all was still. The Cube had vanished, and I was in my cell, yet . . . everything was different, everything was as if seen in a glass. Dizzy and confused, I slept and dreamt of Lineland.

In his dream of Lineland, A Square imagines that once again he is talking to the king of Lineland, a segment with a bass voice at his left end and a tenor voice at his right end. Seized with a desire to really upset the king, A Square reaches down into Lineland and flips the king over, rotating him around his central point. The other Linelanders can hear that the king doesn't sound right: he's been turned into his own mirror image. In a frenzy, they set upon the king and tear him to shreds.

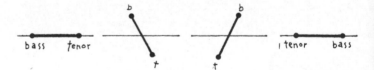

Fig. 45. The king of Lineland flips.

When A Square wakes up the next morning, everything still looks backwards. And the guard who brings breakfast takes one look at our friend and begins to scream. It's really true, that crazy Cube has turned A Square over, rotating him about his central axis.

Normally, if a Flatlander's eye is on his north side, then his mouth is facing east. But now A Square is the opposite. Just a totally flipped square.

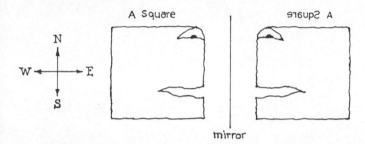

Fig. 46. Flatland mirror.

The Council met, pronounced A Square an "object of horror to the gods," and decided to execute him.

We'll get back to A Square's adventures in a while, but now let's stop and ask what it would be like to get "turned over" in the fourth dimension. It would seem that a 4-D being could turn you into your own mirror image by rotating you, in the fourth dimension, around a plane that cuts through your body — say the plane that includes the tip of your nose, your navel, and your spine. That one plane of your body would stay in our space. Your right half would move *ana*, let us say, and your left half would move *kata*. The two halves would, in their parallel spaces, move past the plane of rotation, and then they would swing back into our space. Rotation about a plane is hard for us to imagine . . . But just remember how hard it is for a Flatlander to think of rotation about a line.

PUZZLE 4.1

A cube that intersects a plane at right angles makes a square cross section. Would it be possible to place a cube so that it intersects a plane in a triangular cross section? How? What other cross-sectional shapes could a cube show?

While you were in the process of rotating you would really look strange, for all of you that would remain in our space would be a cross section something like a microtomed tissue. If you were moved up and down at right angles to our space, we could see each of your cross sections in turn, and really get a good 3-D knowledge of your inner workings.

Fig. 47. Before and after.

Actually the diagnostic tool known as a CAT scan consists of a process something like this: the building up of a 3-D model of the body by looking at a series of cross-sectional x-rays.

Mysterious as a four-dimensional rotation into one's own mirror image may seem, it is actually possible for us to watch such a rotation taking place, as we will see shortly. The method, which provides genuine insight into the fourth dimension, has to do with A Cube.

Fig. 48. A Cube.

A Cube, you will notice, has typically schizoid features. The right half of his face, which is controlled by the analytic and sociable left brain, has a friendly smile and a precise triangular eye. The left half of his face, controlled by the dark and intuitive right brain, has a lax expression and a wandering round eye. Clearly A Cube and A Cube's mirror image are totally different. A Cube has his triangle eye on the right, but ǝdu⅃ A has his triangle eye on the left.

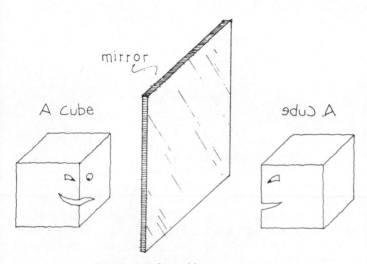

Fig. 49. A Cube and his mirror image.

Just as no amount of sliding around in the plane ever could have turned A Square into his own mirror image, no 3-D contortions could ever turn A Cube into ǝdu⅃ A. But, as the mathematician August Ferdinand Möbius discovered in 1827, it is in fact possible to turn a 3-D object into its mirror image by an appropriate rotation through four-dimensional space.

Surprisingly enough, it is actually possible for our minds to perform such a rotation. Many readers will be familiar with the "reversing cube diagram," also known as the Necker cube. If one stares at a skeleton drawing of a cube for a while, the mind's 3-D interpretation of the figure flip-flops back and forth between the two versions shown. If you have trouble getting the figure to "do" this, it might help to focus

It is to be regretted that Plattner's aversion to the idea of postmortem dissection may postpone, perhaps forever, the positive proof that his entire body has had its left and right sides transposed. Upon that fact mainly the credibility of his story hangs. There is no way of taking a man and moving him about *in space*, as ordinary people understand space, that will result in our changing his sides. Whatever you do, his right is still his right, his left his left. You can do that with a perfectly thin and flat thing, of course. If you were to cut a figure out of paper, any figure with a right and left side, you could change its sides simply by lifting it up and turning it over. But with a solid it is different. Mathematical theorists tell us that the only way in which the right and left sides of a solid body can be changed is by taking that body clean out of space as we know it — taking it out of ordinary existence, that is — and turning it somewhere outside space. This is a little abstruse, no doubt, but anyone with a slight knowledge of mathematical theory will assure the reader of its truth. To put the thing in technical language, the curious inversion of Plattner's right and left sides is proof that he has moved out of our space into what is called the Fourth Dimension, and that he has returned again to our world. Unless we choose to consider ourselves the victims of an elaborate and motiveless fabrication, we are almost bound to believe that this has occurred.

H. G. WELLS,
"The Plattner Story," 1896

your attention near the middle of the drawing and try mentally "pushing" or "pulling" at one of the corners.

What makes the Necker cube reversal so important here is that the two possible 3-D interpretations of the original skeleton drawing are in fact mirror images of each other!

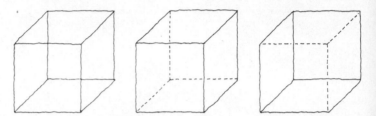

Fig. 50. The Necker cube and its two interpretations.

We can see this if we draw A Cube's features in on the Necker cube, and imagine that A Cube's body is clear as glass. If we take the features to be on the side of the cube closer to us, then we are looking just at A Cube, who has a triangle eye on his right. But if we interpret the features as being on the side of the cube further away from us, then we are looking at a rear view of əduꓳ A, A Cube's mirror image, who has a triangle eye on his *left*. If A Cube is transparent, then the original ambiguous figure can be seen as either the Cube or the Cube's mirror image.

The point of this is that the sort of twinkling rearrangement that takes place when a Necker cube reverses is equivalent to a rotation through the fourth dimension. It happens so fast it's hard to catch, but if you watch it for a

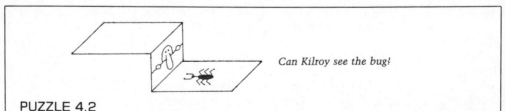

Can Kilroy see the bug?

PUZZLE 4.2

Here is a very remarkable Necker-type illusion. Can the little man see the beetle or not?

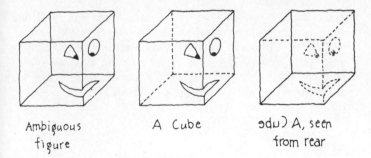

Ambiguous figure　　　A Cube　　　ɘduƆ A, seen from rear

Fig. 51. A Cube changing into ɘduƆ A.

while it begins to feel like maybe you're getting a peek into the fourth dimension. The hypercube drawing, by the way, is particularly good to look at in terms of Necker cube reversals. Stare at it as at a mandala and the thing fairly seethes with activity, doing its best to get hyper.

Paradise? I experienced all of that recently.

Paradise must consist of the stopping of pain. That means, however, that we live in Paradise as long as we have no pain! And we don't even know it.

Happy and unhappy people live in the same world, and they don't even know it!

I have the feeling as if during the past months I have been walking around my own life in a fantastic, mysterious maze, and now I have returned precisely to that spot at which I began. But, since I moved outside the normal dimensions, right and left somehow got exchanged. My right hand is now my left one, my left hand my right one.

Returned into the same world and see it now as a happy one.

The shreds of peeled paint on the door belong to a mysterious work of art.

LARS GUSTAFSSON,
The Death of a Beekeeper, 1978

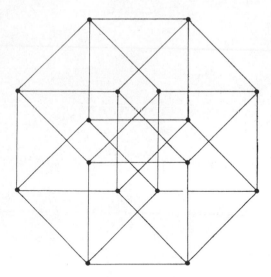

Fig. 52. The Hypercube.

An especially dramatic version of the Necker cube can be created by folding together the Neck-A-Cube model depicted in figure 53. When it is folded, the paper is in the shape of Ǝq∩Ɔ A, as seen from the inside of his hollowed-out head. After Necker reversal pulls the corner out at you, it looks like A Cube should. But when you break the illusion (by opening your other eye, or by turning the model far enough to show its back), then the thing "topples" through the fourth dimension back to the hollowed-out Ǝqn A form.

1. Trace figure.
2. Cut out around outline.
3. Crease line *AC*, and then crease line *DE*. Each time crease by folding the marked surfaces together.
4. Slit from *A* to *B*.
5. Slide one of the upper "squares" behind the other to make something like the corner of a room where walls meet ceiling. Cup the object in your right hand.
6. Close one eye and stare at the corner. "Pull" at the corner till Necker cube reversal takes place.

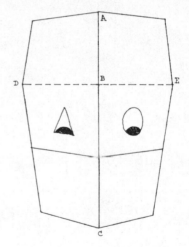

Fig. 53. Design for the Neck-A-Cube.

7. Once the object is solidly reversed, try moving your hand around.
8. If you have trouble getting the illusion, make sure that the model is uniformly lit (so that shadows don't provide depth cues); and make sure to hold it still till Necker reversal has taken place.

Once the cube is reversed, it seems to move in an un-natural way. If you have it reversed and cradle it in your hand, then it will seem to stick up in the air and rock in the opposite way that your hand moves. An even more remark-able effect is obtained if you set the cube down, or glue it to the wall, and then reverse it. Now, if you can manage to hold it reversed as you move your head around, you will see the cube apparently turning so as to always stare right at you!

I should mention here that my model is based on a similar design by Jerry Andrus, a brilliant illusionist living in Al-bany, Oregon. Andrus calls his reversing pattern a Parabox.

Staring at the Neck-A-Cube too much before bedtime can lead to really unpleasant dreams. Although the fourth di-mension is of great theoretical and philosophical interest, there is something frightening and mind-warping about it when it starts to get too real. I remember in particular a series of dreams I had in 1976, soon after I came to under-stand that left and right are concepts as relative as up and down, or front and back.

In these dreams I would be walking down a street with someone on either side of me — my wife, Sylvia, let us say, on my left, and my friend Greg on my right. I would move out of my body and watch the three of us from a distance, first from a point in our space, then from a point entirely outside of space. What struck me was that *depending on which half of hyperspace I looked from*, the order of the three people would be Sylvia-Rudy-Greg, or Greg-Rudy-Sylvia. In my dream of walking down the street, I would begin to internalize such a shift, and the whole city around me would change back and forth between itself and its own mirror image. Some mornings I would wake up convinced that I and the whole universe had been replaced by our mir-ror images overnight. The worst day of all was when I sat up in bed and actually saw the reversal taking place: the whole room, furniture and all, rambunctiously do-si-do-ing about

"In actual, physical life I can turn as simply and swiftly as anyone. But men-tally, with my eyes closed and my body immobile, I am unable to switch from one direction to the other. Some swivel cell in my brain does not work. I can cheat, of course, by setting aside the mental snapshot of one vista and leisurely selecting the opposite view for my walk back to my starting point. But if I do not cheat, some kind of atrocious obstacle, which would drive me mad if I persevered, prevents me from imagining the twist which transforms one direc-tion into another, directly opposite. I am crushed, I am carrying the whole world on my back in the process of trying to visualize my turn-ing around and making my-self see in terms of 'right' what I saw in terms of 'left' and vice versa."

VLADIMIR NABOKOV,
Look at the Harlequins, 1974

PUZZLE 4.3

Looking at a Necker cube can help us to think about the fourth dimension. Can you think of a "Flatland Necker Illusion" that A Square could use to help think of the third dimen-sion?

into mirror reversal. But, of course, I couldn't prove anything . . . I'd turned over, too.

After these experiences I often wondered if there could be any objective fact correlated to a person's feeling that the world has turned into its mirror image. One possible idea that came to mind was that a person might have a 4-D component of mind that projects out of the brain and into hyperspace. If this "hyperbrain" material were to shift from being primarily on, let us say, the *ana* side to being primarily on the *kata* side, then one would be justified in saying that the world has changed — in the same way that the palm drawn in figure 42 changes when an observer moves from one side to the other.

Sometimes I get so confused about left and right that all I'm really sure about is what is next to what. I get this feeling if I go to England . . . or just watch an English movie . . . where everyone drives on the left side of the road. A very commonplace instance of left-right confusion arises if you happen to study your teeth in the mirror while feeling them with your tongue. Left and right can get hopelessly mixed up in short order. Charles Hinton, whom we will talk about in the next chapter, felt that deliberately inducing such a confusion about left and right was the best path to four-dimensional thought.

5

Ghosts from Hyperspace?

SPIRITUALISM — the belief that spirits of the dead are nearby and eager to contact us — has never been so popular as in the late nineteenth century. Throughout the United States, England, and Europe, amateur or professional mediums were organizing séances. A group of people would sit around a table in a darkened room, the medium would moan and mutter, and then the spirits would manifest themselves.

And what would said spirits do? They would make noises: sharp raps on the table. They would move things: turning or rocking the table and, occasionally, even lifting the table up into the air. They would send messages: often by causing a piece of pencil to scratch some words on a slate held under the table. They would materialize as white fogs, or sometimes as spirit hands that would reach up over the table's edge.

Spiritualism mediums were, of course, under heavy suspicion of fraud from the very beginning. People are so eager to believe in life after death that their reports of what they experience at a séance are not very reliable. No accumulation of colorful anecdotes can ever take the place of a really scientific investigation. The few scientists who believed in spiritualism began trying to find some kind of solid theory to support their belief in ghosts.

Quite abstractly, there seem to be two possibilities if

Fig. 54. Occult forces.

(1) Physically, the souls of the dead come into the thraldom of certain living beings who are called mediums. These mediums are, for the present at least, a not widely diffused class, and they appear to be almost exclusively Americans. At the command of these mediums, departed souls perform mechanical feats which possess throughout the character of absolute aimlessness. They rap, they lift tables and chairs, they move beds, they play on the harmonica, and do other similar things. (2) Intellectually, the souls of the dead enter a condition which, if we are to judge from the productions which they deposit on the slates of the mediums, must be termed a very lamentable one. These slate-writings belong throughout in the category of imbecility; they are totally bereft of any contents. (3) The most favored, apparently, is the moral condition of the soul. According to the testimony which we have, its character cannot be said to be anything else than that of harmlessness. From brutal performances, such, for instance, as the destruction of bed-canopies, the spirits most politely refrain.

WILHELM WUNDT,
"Spiritualism, A Question of Science So-Called," 1889

spirits exist. Either they are in our space, but very insubstantial, or they are somehow outside our space altogether. The idea of spirits as tenuous forms in our space was popular with some early spiritualists: they held that the spirits were made of "ectoplasm" or, even less concretely, of "vibrational energy." One difficulty here was that if the spirits were so very insubstantial, how could they do things like lifting a heavy séance table?

No such difficulty arises if we think of spirits as solid and substantial beings, but here of course we have the question of why one doesn't notice spirits very often if they really are so solid. The answer to this is to assert that the home of spirits is somehow outside our space entirely. How best to have spirits live outside space? One could have them be infinitely far away, but then one would have the problem of how they can get here so rapidly when summoned up by a medium. A much more satisfactory explanation is to say that *spirits live in the fourth dimension.*

The beauty of this idea is that the spirits are wholly outside of our normal, material space yet they are . . . at the same time . . . right next to us, waiting just a few feet *ana* or *kata*. Although the idea of spirits as 4-D beings had its greatest popularity in the nineteenth century, it had been hinted at some two hundred years before, by the Cambridge Platonist Henry More (1614–1687). Like the scientific spiritualists, More was opposed to the idea that spirits, angels, and Platonic forms could exist as insubstantial abstractions. He felt that if spirits really exist, then they must actually *take up space.* Yet, if a person's soul takes up space, we have the question of how it can fit into the person's solid physical body. In 1671, More came up with the suggestion that spirits should be four-dimensional. He phrased this in terms of an occult quality he called *spissitude*, meaning something like "denseness of substance." His idea seems to have been that the differences between the physically identical bodies of a dead person and a living person would be that the living body has more spissitude, and that spissitude is physically unobservable because it corresponds to a certain hyperthickness in the direction of the fourth dimension.

The person who really popularized the spiritualist notion of ghosts from the fourth dimension was Johann Carl Friederich Zöllner (1834–1882). Zöllner was a professor of astronomy at the University of Leipzig, the same university where August Möbius made his 1827 discovery that it would be possible to turn a 3-D object into its mirror image by a hyperspace rotation, and the same university where Gustav Fechner wrote his 1846 essay "Why Space Has Four Dimensions." Zöllner got his interest in spiritualism from an 1875 trip to England, where he visited William Crookes, inventor of the cathode-ray tube.

Crookes was very committed to spiritualism, and was the champion of the American medium Henry Slade. When Slade's stay in England ended in arrest and conviction for fraud, the medium went to visit Zöllner, who was eagerly waiting for someone to help him prove that spirits are four-dimensional. According to Zöllner's *Transcendental Physics* of 1878, the experiments were an immediate success.

The first thing Slade did was to tie four simple overhand knots in a string Zöllner provided. What made this feat so striking was that the string was originally unknotted, with its ends joined together by a glob of sealing wax impressed with Zöllner's own seal. Of course, in reality Slade undoubtedly managed to switch the strings, but if he had not cheated, his trick would have been genuinely four-dimensional.

Why? Well, if a spirit could move a little segment of the string *ana* out of our space, then it would be, for all practical purposes, like having a gap in the string so that one could move it "through" itself to get a knot. Once the string is in the proper position, you move the displaced segment back *kata* to our space and you've tied a knot without moving the ends of the string.

That's one way to do it. An easier way is to tie the knot

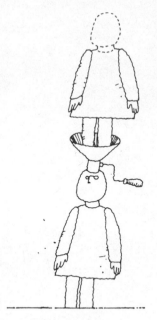

Fig. 55. Body and soul.

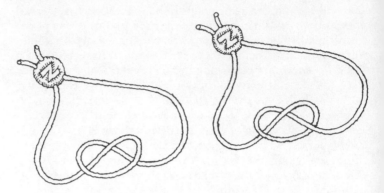

Fig. 56. Zöllner's string, before and after.

first and then seal the ends together. In and of itself, a knotted string with its ends sealed together does not immediately make one believe that the knot was tied by a four-dimensional spirit. Zöllner was of course aware of this, and he designed several interesting tests by which Slade's spirit friends could have established lasting proof of their four-dimensionality. Three of the tests are described in *Transcendental Physics*:

1. Two wooden rings, one of oak, the other of alderwood, were each turned from one piece . . . Could these two rings be interlinked without breaking, the test would be additionally convincing by close microscopic examination of the unbroken continuity of the fibre. Two different kinds of wood being chosen, the possibility of cutting both rings from the same piece is likewise excluded. Two such interlinked rings would consequently in themselves represent a "miracle," that is, a phenomenon which our conceptions heretofore of physical and organic processes would be absolutely incompetent to explain.

2. Since among products of nature, the disposition of whose parts is according to a particular direction, as with snail-shells twisted right or left, this disposition can be reversed by a four-dimensional twisting of the object, I had provided myself with a large number of such shells, of different species, and at least two of each kind.

3. From a dried gut, such as is used in twine-factories, a band without ends was cut. Should a knot be tied in this band, close microscopic examination would also reveal whether the connection of the parts of this strip had been severed or not.

So the idea was that Slade's spirits should link the two wood rings, turn some snail shells into their mirror-image forms, and produce a knot in a closed loop of skin cut from a pig's bladder. Did it work?

> Seldom happens just that which we, according to the measure of our limited understanding, wish; but if, looking back on the course of some years, we regard what has actually come to pass, we recognize gratefully the intellectual superiority of that Hand which, according to a sensible plan, conducts our fates to the true welfare of our moral nature, and shapes our life dramatically to a harmonic whole.

In other words, no. Instead of doing what Zöllner had wanted, the spirits put the rings around a table leg, moved a snail shell from the tabletop to the floor beneath the table, and burned a spot on the bladder band.

Very few scientists were convinced by Zöllner's experiments. Although Zöllner himself may well have been an honest man, he was almost unbelievably gullible — an unworldly scientist easily taken in by a professional charlatan like Slade. One might think that now, a century later, scientists could no longer be fooled by cheap conjuring tricks, but this does not seem to be the case. Just a few years ago, Uri Geller — who is almost certainly a fraud — was able to obtain the backing and endorsement of a number of physicists at the Stanford Research Institute. A glance at any supermarket tabloid makes it clear that the public's interest in ghosts is greater than it has ever been.

Fig. 57. Evidence of 4-D ghosts! (Engraving from J. C. F. Zöllner, Transcendental Physics.)

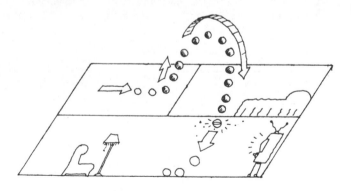

Fig. 58. Where the balls went in Poltergeist.

Most modern ghost tales do not bring in the fourth dimension. An interesting exception is Steven Spielberg's movie *Poltergeist.* The 4-D aspect of this movie arises when balls that are thrown into the closet of one room appear from the ceiling of another room . . . indicating a route through the fourth dimension!

The main effect of Zöllner's work was that the fourth dimension began to seem disreputable and unscientific. Yet his basic message, *spirits live in the fourth dimension*, did not fall on totally deaf ears. The notion of beings living in an unseen hyperspace world was taken up by turn-of-the-century Protestant clergymen all over England.

The best known of these clergymen is, of course, Edwin Abbott, but there were many others who enthusiastically adopted the notion that heaven, hell, our souls, the angels, and God himself could be comfortably lodged in some higher dimension. The basic credo of these Christian spiritualists can be found in A. T. Schofield's 1888 book, *Another World*:

We conclude, therefore, that a higher world than ours is not only conceivably possible, but probable; secondly, that such a world may be considered as a world of four dimensions; and thirdly, that the spiritual world agrees largely in its mysterious laws, in its language which is foolishness to us, in its miraculous appearances and interpositions, in its high and lofty claims of omniscience, omnividence, etc., and in other particulars, with what by analogy would be the laws, language, and claims of a fourth dimension . . .

Though the glorious material universe extends beyond the utmost limits of our vision, even artificially aided by the most

Fig. 59. "Aided by the most powerful telescopes."

powerful telescopes, that does not prevent the spiritual world and its beings, and heaven and hell being by our very side.

Far from these spiritual regions occupying some small corner of the material universe, as surely as the greater includes the less, so surely is the material universe, vast as it is, swallowed up in the spiritual.

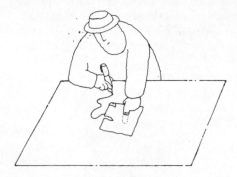

Fig. 60. God the Father?

To a sophisticated thinker, there is something just a little too crude and materialistic about having souls and angels and God be literal chunks of hypermatter in four-dimensional space. Why precisely *four*? If any kind of higher-space view of higher reality is reasonable, I would think it to be the notion that our souls are patterns in *infinite*-dimensional Hilbert space. A second objection to the idea of spirits and angels swarming about our space is the feeling that, at higher levels, the whole concept of separately acting individuals might well fall away.

Let me illustrate this last thought with a Flatland picture, figure 61.

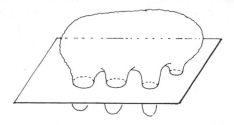

Fig. 61. The One and the Many.

What seems to be a collection of many individuals may turn out, at some higher level, to just be parts of one higher entity. There is something lacking in a world view that leaves out the notion of higher unity and has diverse spirits poking at our world like pool sharks at a game of eight ball.

Another failing in spiritualism, Christian or not, is the reliance on anecdotes about miracles and psychic feats. If higher dimensions have any real meaning for us, that meaning should be part of our ordinary life. I don't believe in miracles; I don't disbelieve in them. The whole subject just leaves me cold. You can walk on water? You can bend spoons? So what. You've proved that nobody really walks on water? You've proved that nobody really bends spoons? I couldn't care less. What I want to know is much simpler: *What is it like to be alive?*

Fig. 62. Bend that spoon.

"We live in a three-dimensional space" is really quite a complex statement, and not a statement that every person would automatically think of as true. Explaining to someone why it is that space is three-dimensional really involves convincing them that they should think of the world around them in a certain way. Imagine an alien culture whose architecture is not based on cubes. Without the inevitable three perpendicular lines in each corner of one's room, the notion of reducing all experiences to the oscillations of a three-knobbed Etch-A-Sketch might seem a bit strange. What about feelings, what about levels of thought, what about dreams?

What I am trying to suggest here is that perhaps we ourselves are, in some very real sense, beings of more than three dimensions. P. D. Ouspensky wrote something very interesting about this in his 1908 essay "The Fourth Dimension."

If the fourth dimension exists, one of two things is possible. Either we ourselves possess the fourth dimension, i.e., are beings of four dimensions, or we possess only three dimensions and in that case do not exist at all.

If the fourth dimension exists while we possess only three, it means that we have no real existence, that we exist only in somebody's imagination and that all our thoughts, feelings and experiences take place in the mind of some other higher being, who visualizes us. We are but products of his mind, and the whole of our universe is but an artificial world created by his fantasy.

If we do not want to agree with this we must recognize ourselves as beings of four dimensions.

Do we not in sleep live in a fantastic fairy kingdom where everything is capable of transformation, where there is no stability belonging to the physical world, where one man can become another or two men at the same time, where the most improbable things look simple and natural, where events often occur in inverse order, from end to beginning, where we see the symbolical images of ideas and moods, where we talk with the dead, fly in the air, pass through walls, are drowned or burnt, die and remain alive?

All this taken together shows us that we have no need to think that the spirits that appear or fail to appear at spiritualistic séances must be the only possible beings of four dimensions. We may have very good reason for saying that we are ourselves beings of four dimensions and we are turned towards the third dimension with only one of our sides, i.e., with only a small part of our being. Only this part of us lives in three dimensions, and we are conscious only of this part as our body. The greater part of our being lives in the fourth dimension, but we are unconscious of this greater part of ourselves. Or it would be still more true to say that we live in a four-dimensional world, but are conscious of ourselves only in a three-dimensional world. This means that we live in one kind of condition, but imagine ourselves to be in another.

For Ouspensky, the fourth dimension was not only a spatial concept but a type of consciousness, an awareness of greater complexities and higher unities. For him, the mathematical study of the fourth dimension led quite naturally to a belief in the teachings of mysticism. In their simplest form, these teachings are but two: *All is One* and *The One is Unknowable.*

The music faded into silence as the grotto became completely dark except for the brilliantly illuminated front wall. The shadow of the minister rose before us. After announcing the text as Ephesians, Chapter 3, verses 17 and 18, he began to read in low, resonant tones that seemed to come directly from the shadow's head: "... that ye, being rooted and grounded in love, may be able to comprehend with all saints what is the breadth, and length, and depth, and height ..."

It was too dark for note-taking, but the following paragraphs summarize accurately, I think, the burden of Slade's remarkable sermon.

Our cosmos — the world we see, hear, feel — is the three-dimensional "surface" of a vast, four-dimensional sea ...

What lies outside the sea's surface? The wholly other world of God! No longer is theology embarrassed by the contradiction between God's immanence and transcendence. Hyperspace touches every point of three-space. God is closer to us than our breathing. He can see every portion of our world, touch every particle without moving a finger through our space. Yet the Kingdom of God is completely "outside" of three-space, in a direction in which we cannot even point.

MARTIN GARDNER, "The Church of the Fourth Dimension," 1962

Fig. 63. All is One.

What, you may ask, do these two high-sounding precepts have to do with the fourth dimension? Quite briefly, the first idea is that higher space can be viewed as a background of connective tissue tying together the world's diverse phenomena. If one moves toward higher and higher conceptions of space, one is tending toward some ideal "Superspace" in which everything — near and far, past and future, big and small, real and imagined — is together in some great Unity.

In his 1885 essay "Many Dimensions," the great hyperspace philosopher Charles H. Hinton speaks very clearly and colorfully about his identification of the abstract idea of "space" with the "One" of mysticism:

> Yet with them [the Eastern mystics] I feel an inward sympathy, for I too, as they, have an inward communion and delight, with a source lying above all points and turns and proofs — an inward companion, whose presence in my mind for one half-hour is worth more to me than all the cosmogonies that I have ever heard of, and of which all the thoughts I have ever thought are but minutest fragments, mixed up with ignorance and error. What their [the mystics'] secret is I know not, mine is humble enough — the inward apprehension of space.
>
> And I have often thought, travelling by railway, when between the dark underground stations the lads and errand boys bend over the scraps of badly printed paper, reading fearful tales — I have often thought how much better it would be if they were doing that which I may call "communing with space." 'Twould be of infinite delight, romance, and interest; far more than are those creased tawdry papers, with no form in themselves or in their contents.

And yet, looking at the same printed papers, being curious, and looking deeper and deeper into them with a microscope, I have seen that in splodgy ink stroke and dull fibrous texture, each part was definite, exact, absolutely so far and no farther, punctiliously correct; and deeper and deeper lying a wealth of form, a rich variety and amplitude of shapes, that in a moment leapt higher than my wildest dreams could conceive.

And then I have felt as one would do if the dark waters of a manufacturing town were suddenly to part, and from them, in them, and through them, were to uprise Aphrodite, radiant, undimmed, flashing her way to the blue beyond the smoke; for there, in these crabbed marks and crumpled paper, there, if you but look, is space herself, in all her infinite determinations of form.

So for Hinton, space itself is the medium by which one is to grasp the world's unity. And what of the second teaching of mysticism, *The One is Unknowable*? Here we must be clear on the sense in which the word "know" is to be used. Mystically speaking, the One *is* knowable, knowable in the sense that we can feel space around us, knowable in the sense that we can open our hearts to feel life and beauty and love. It is only for the *rational* mind that the One is unknowable.

No finite string of human symbols can really represent the ultimate reality: call it God, the One, the Absolute, Everything, or what you will. The situation here is a bit analogous to trying to understand an infinite set such as the set N of all whole numbers $(1, 2, 3, \ldots)$. Given the idea of number, we do have a pretty good understanding of what N is — this higher understanding is to be likened to the mys-

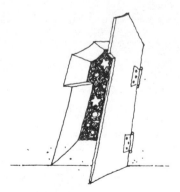

Fig. 64. The One is Unknowable.

tical knowing of the One. But if we insist on a complete, explicit listing of members, then the set N will always lie beyond our grasp — this is comparable to the rational mind's inability to fully grasp the One.

In the following long excerpt (also from "Many Dimensions"), Hinton makes clear the sense in which space is, at the deepest level, to be known by the heart and not by the brain.

And, indeed, space is wonderful. We all know that space is infinite in magnitude — stretching on endlessly.

And when we look quietly at space, she shows us at once that she has infinite dimensions.

And yet, both in magnitudes and dimensions there is something artificial.

To measure, we must begin somewhere, but in space there is no "somewhere" marked out for us to begin at. This measuring is something, after all, foreign to space, introduced by us for our convenience.

And as to dimensions, in order to enumerate and realize the different dimensions, we must fix on a particular line to begin with, and then draw other lines at right angles to this one.

But the first straight line we take can be drawn in an infinite number of directions. Why should we take any particular one?

If we take any particular line, we do something arbitrary, of our own will and decision, not given to us naturally by space.

No wonder then if we take such a course we are committed to an endless task.

We feel that all these efforts, necessary as they are to us to apprehend space, have nothing to do with space herself. We introduce something of our own, and are lost in the complexities which this brings about.

May we not compare ourselves to those Egyptian priests who, worshipping a veiled divinity, laid on her and wrapped her about with ever richer garments, and decked her with fairer raiment.

So we wrap around space our garments of magnitude and vesture of many dimensions.

Till suddenly, to us as to them, as with a forward tilt of the shoulders, the divinity moves, and the raiment and robes fall to the ground, leaving the divinity herself, revealed, but invisible; not seen, but somehow felt to be there.

And these are not empty words. For the one space which is not this form or that form, not this figure or that figure, but which is to be known by us whenever we regard the least details of the visible world — this space can be apprehended. It is not the shapes and things we know, but space is to be apprehended in them.

Fig. 65. *"The least details of the visible world."*

The true apprehension and worship of space lies in the grasp of varied details of shape and form, all of which, in their exactness and precision, pass into the one great apprehension.

And we must remember that this apprehension does not lie in the talking about it. It cannot be conveyed in description.

We must beware of the attitude of standing open-mouthèd just because there is so much mechanics which we do not understand, but geometry and mathematics only spring up there where we, in our imperfect way, introducing our own limitations, tend towards the knowledge of inscrutable nature.

If we want to pass on and on till magnitude and dimensions disappear, is it not done for us already? That reality where magnitude and dimensions are not, is simple and about us. For passing thus on and on we lose ourselves, but find the clue again in

Fig. 66. *"The simplest acts of human goodness."*

the apprehension of the simplest acts of human goodness, in the most rudimentary recognition of another human soul wherein is neither magnitude nor dimension, and yet all is real.

The author of this strange yet compelling passage led — in the context of his time — an equally strange life.

Charles Howard Hinton, known to his family as Howard, was born in London in 1853. His father, James, was a well-known author who over the course of his life moved from being an ear surgeon to a religious philosopher to the champion of a new sexual morality. In his declining years, James Hinton managed to surround himself with a circle of admiring women, most of whom were sexually intimate with him. One of his sayings was, "Christ was the Saviour of men, but I am the saviour of women, and I don't envy Him a bit!"

Howard obtained his B.A. from Oxford in 1877, and soon after he married Mary Boole, daughter of the famous logician who invented "Boolean algebra." In 1880 Hinton obtained a post as science master at the Uppingham School, meanwhile continuing work for a master's degree in mathematics.

Despite all of his education, Hinton felt somehow adrift, as if he could not get hold of any really concrete knowledge. For who knows what reason, he hit on the idea of memorizing a cubic yard of one-inch cubes. That is, he took a 36-by-36-by-36 block of cubes, assigned a two-word Latin name (for example, *Glans Frenum*) to each of the 46,656 units, and learned to use this network as a sort of "solid paper." Thus, when he wished to visualize some solid structure, he would do so by adjusting its size so that it fit into his cubic yard. Then he could describe the structure by reviewing the list of occupied cells. (Although this sounds incredible, it is not really impossible — Hinton had a system that reduces the number of brute facts to 216.)

As it turned out, this seemingly insane idea was a fantastic source of inspiration for Hinton. For what he had in effect done was to create within his mind the kind of "three-dimensional retina" that a 4-D being would have (as we discussed in chapter 3). Some mysterious inspiration drove Hinton forward on the correct path, and he next had the idea of learning his block of cubes in each of its twenty-four possible orientations (six choices for the bottom face times four choices for the front face).

The result of this was that Hinton was now able to take a 3-D object and see its parts simply in terms of what is next to what, thus freeing himself from our spacebound concepts of front-back and up-down. Shortly after this, he took the final step and let the concept of left-right fall away, thus arriving at a perfectly four-dimensional view of the world (as we discussed in chapter 4). Now Hinton could, without difficulty, visualize all the cross sections of a hypercube, undisturbed by the way in which these cross sections can become their own mirror images depending on where the hypercube lies relative to our space.

As his understanding of the fourth dimension grew, Hinton set to writing about it. His first published essay, "What Is the Fourth Dimension?," appeared in 1880 in the *Dublin University Magazine*, was reprinted in the *Cheltenham Ladies' College Magazine* of September 1883, and finally was released as a pamphlet, with the crassly commercial subtitle "Ghosts Explained," by Swann Sonnenschein & Co. in 1884. In the period 1884–1896, Swann Sonnenschein published nine pamphlets by Hinton, which were then collected in the two-volume set *Scientific Romances*. Some of the pieces were essays on the fourth dimension, and some were what we would now call science fiction stories.

Hinton might have settled into a comfortable life like his father's: the life of a writer respected by intellectuals and loved by many women. But in 1885 disaster struck. Charles

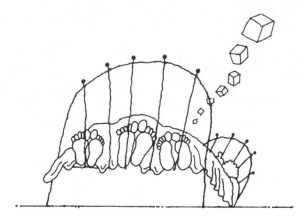

Fig. 67. Hinton was a bigamist.

Howard Hinton was arrested for bigamously marrying a woman called Maude Weldon, with whom he had spent a week in King's Cross Hotel, and by whom he was the father of twins. The headmaster at Uppingham had, till this time, been under the impression that Maude Weldon was Hinton's sister — apparently she was a regular guest of Howard and his legitimate wife, Mary. Of course, he lost his job. His career was ruined. When he came up for sentencing in 1886 the magistrate let him off with a token three days in jail. Soon after this, Howard, Mary, and their children fled to a teaching post at a middle school in Yokohama, Japan.

The torment Hinton must have undergone during this period shows in two of his "Scientific Romances": "Stella" and "An Unfinished Communication." "Stella" tells of a girl whose guardian has made her invisible, so that she will not fall into the traps of vanity and physical love. Yet Stella does find a lover, a man who forces her to wear clothes and make-up so that all can see her. It is easy to imagine that Maude Weldon is the model for Stella.

"An Unfinished Communication" is a very strange story indeed. It tells of a desperate man who seeks the services of an Unlearner, a man who can make him forget his horrible past. The Unlearner stresses the importance of living openly and sharing one's secrets, something Hinton must have been frantic to do as he sank deeper and deeper into the morass created by his attempts to live out his father's free-love philosophy.

When Hinton left for Japan, he left a manuscript in the hands of two friends. This book, *A New Era of Thought*, appeared in 1888. Here Hinton outlines in detail his system

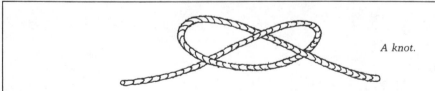

A knot.

PUZZLE 5.1

A line or string can only be knotted in 3-D space: no string can be knotted in 2-D space, and no knot will stay tied in 4-D space. Why not? In 4-D space it is possible to knot a plane. Can you imagine how?

for learning to think four-dimensionally by manipulating a set of eighty-one colored cubes, cubes that are to stand for the eighty-one parts of a 3-by-3-by-3-by-3 hypercube. Hinton is at his most confident here, and he asserts that "the particular problem, at which I have worked for more than ten years, has been completely solved. It is possible for the mind to acquire a conception of higher space as adequate as that of our three-dimensional space, and to use it in the same manner."

In *A New Era of Thought* Hinton also explains *why* it should be possible for us to form 4-D thoughts. People sometimes feel that since our brains are made up of 3-D nets of neurons it is futile to hope to model 4-D patterns in our brains. But, replies Hinton, is it not possible that our space has some slight 4-D hyperthickness? If this is indeed the case, then the small bits of matter that code up our thoughts may be free to form actual 4-D patterns, and 4-D thought is easily achieved.

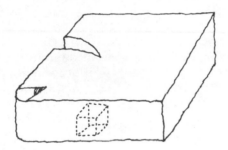

Fig. 68. Thick Square with mental image of A Cube.

It was at one time possible to purchase sets of Hinton cubes from his publisher. I once made myself a set of them, and was somewhat disappointed. Playing with the colored cubes may not really be the best way to start thinking four-dimensionally.

Eighteen ninety-three found Howard Hinton at Princeton University, as an instructor in the Mathematics Department. Oddly enough, Hinton devoted most of his energies while at Princeton to the invention of a "baseball gun." The point of having such a gun was that during the week's practice sessions the Princeton team could practice hitting

Although on account of his enthusiasm for metageometry he was never a great success as an instructor in his college positions, he made many friends, and in Princeton endeared himself to the students by one of the most successful practical jokes ever perpetrated there. This was just after his perfection of the baseball gun. He invited the faculty and students to a lecture at which he demonstrated the machine and described its scientific theory. While he was upon the platform the lecture was interrupted by the arrival of a special delivery postman who walked down the aisle and called to the professor. As he had been the victim of many practical jokes and "horsed" by the students according to the Princeton custom, the audience prepared for some absurd diversion. After protesting against the interruption, but not being able to send away the messenger, Prof. Hinton begged permission to look at a letter important enough to demand consideration at such a time. He read aloud, and had turned two pages, reading an account of a baseball game in the year *1950*, before the students discovered that the joke was upon them.

GELETT BURGESS, "The Late Charles H. Hinton," 1907

really fast balls — without the pitchers having to wear out their arms. The gun could shoot a baseball at speeds of up to seventy miles per hour, and two rubber-coated steel fingers attached to the muzzle made it possible to shoot curves as well.

Hinton was fired before long, and found his next job at the University of Minnesota. In 1900 he left Minnesota for Washington, D.C., where he worked first at the Naval Observatory and then at the Patent Office. His faithful wife Mary followed Howard through all these peregrinations, bearing him four sons. She was known in Washington as a lecturer on poetry.

Hinton died suddenly and dramatically at the age of fifty-four, in 1907. A contemporary news article on his death, headlined SCIENTIST DROPS DEAD, describes how Hinton fell dead after complying with the toastmaster's request for a toast to "female philosophers," at the annual banquet of the Washington, D.C., Society of Philanthropic Inquiry.

Charles Howard Hinton lived a wild and varied life. I would like to think that what sustained him through it all was that mystical vision of space he enjoyed as a young man.

I hope that by now most readers will have some feeling for the fourth dimension as a concept interesting in its own right. In parts II and III of this book we will proceed to see how the fourth dimension can be used to understand the true nature of space, matter, time, and mind.

Part II

SPACE

6

What We're Made Of

WE ARE ACCUSTOMED to thinking of the world as made up of lumps of matter floating in empty space. Matter is something, and space is nothing. But is this really correct? In the past, many powerful thinkers have held that the space between visible objects is filled with a subtler material, a smooth and continuous substance, a "universal plenum," or *aether*.

One does not hear much about the aether these days. But in the nineteenth century the concept was as commonplace as "fields" are now. In his article on aether for the ninth edition of the *Encyclopaedia Britannica*, James Clerk Maxwell (1831–1879), founder of the modern theory of electromagnetism, wrote:

> Whatever difficulties we may have in forming a consistent idea of the constitution of the aether, there can be no doubt that the interplanetary and interstellar spaces are not empty, but are occupied by a material substance or body, which is certainly the largest, and probably the most uniform body of which we have any knowledge.

But why bother filling space up with aether? What's wrong with having space just be the empty background against which matter moves? One problem is that if space is really empty, then it's hard to understand how gravitational force can be transmitted. Isaac Newton set up his famous law of gravity as a quantitative description of the gravitational force, but he was keenly aware that he lacked any understanding of how this force could act across the empty

We have supposed in the case of a plane world that the surface on which the movements take place is inactive, except by its vibrations. It is simply a smooth support.

For the sake of simplicity let us call this smooth surface "the aether" in the case of a plane world.

The aether that we have imagined to be simply a smooth, thin sheet, not possessed of any definite structure, but excited by real disturbances of the matter on it onto vibrations, which carry the effect of these disturbances as light and heat to other portions of matter. Now, it is possible to take an entirely different view of the aether in the case of a plane world.

Let us imagine that, instead of the aether being a smooth sheet serving simply as a support, it is definitely marked and grooved.

CHARLES H. HINTON,
A New Era of Thought, 1888

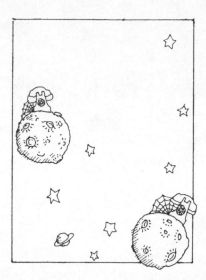

Fig. 69. How do we send signals through the void?

distance of space. Writing in the late seventeenth century, Newton complains:

> It is inconceivable that inanimate brute matter should, without the mediation of something else which is not material, operate upon and affect other matter without mutual contact. That one body can act upon another at a distance, through a vacuum, without the mediation of anything else . . . is to me so great an absurdity, that I believe no man, who has in philosophical matters a competent faculty of thinking, can ever fall into it.

As we will see shortly, the modern theory of gravity embodied in Einstein's general theory of relativity does use a sort of "space-filling aether" to explain how gravitating bodies can influence the motions of faraway objects. For Einstein, the continuous aether is simply space itself, and it is the bending of this aether in higher dimensions that produces gravitational attraction! But I'm getting ahead of myself.

Late in the eighteenth century, Thomas Young and Augustin Fresnel successfully advocated the theory that light is a type of wave motion, rather than a stream of particles. A variety of optical effects, such as diffraction and polarization, seemed to indicate that light must be a vibration in some underlying medium. It was assumed that outer space is filled with a "luminiferous aether," which transmits the wave motion that is light. As the nineteenth century wore on, this aether came to be regarded as the transmitter of magnetic and electrical forces as well.

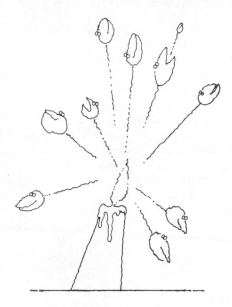

Fig. 70. Wave-particle duality.

Nowadays we are comfortable with the quantum-mechanical idea that light is both particle and wave. A photon is a wavy yet solid little package that can zip through empty space without the benefit of any invisible vibrating jelly underfoot. Just to grasp how alien the notion of a luminiferous aether has become, try to picture a nineteenth-century drawing room from which all the luminiferous aether has been evacuated. The gas jets are burning but no one can see them! A light source with no luminiferous aether to ferry the light's vibrations would be like a bell pounding silently in an evacuated bell jar.

The idea of a room emptied of aether is, of course, quite unrealistic, for most aether theorists felt that ordinary matter offered an open structure to aether. In Thomas Young's words, aether should be able to pass through solid objects "like wind through a grove of trees." If this is the case, then no room's walls could ever keep the aether out.

The idea of an aether wind blowing right through your body is sort of refreshing. Sometimes, standing on the breezy crest of a hill, I have the illusion that the air is blowing through me, dusting off my molecules. If we do think of aether as being the same as space, it would seem to make sense to think of the aether as actually blowing through us as we move through space.

But does this *really* make sense? Certainly we can move relative to other objects, but is space the kind of thing one can move relative to? The answer, as it turns out, is *no*. Sometimes one hears that modern physics has proved there is no aether. What this actually means is that modern physics has proved that it is meaningless to speak of motion relative to empty space. How did this come about?

In the late nineteenth century, physicists and astronomers became interested in determining the Earth's absolute motion. The idea was that space is filled with stagnant, motionless aether, so that as our Earth and our solar system and our galaxy tumble about, we should feel some kind of aether breeze on our faces. The direction in which the breeze feels strongest would then be the direction in which we are really moving.

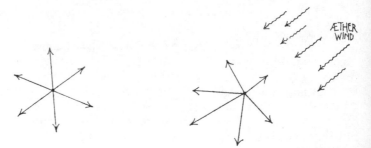

Fig. 71. A light source, with and without aether wind.

"Feeling the breeze" is of course a metaphorical way of speaking: the actual experimental technique was to measure the speed of light in various directions. Light would go fastest when traveling with the aether wind, and slowest when traveling directly against it.

A number of attempts to thus measure the motion of the earth through the aether were made, the most famous of these being the well-known Michelson-Morley experiment of 1887. The results were that a light signal always travels at the same speed, regardless of which direction it goes in. Feeling for the aether wind by measuring speeds of light rays is like sticking a flag out of a spaceship to see which way you're going. Nothing happens: there's no air in outer space, and no wind.

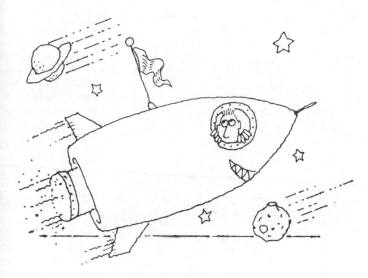

Fig. 72. No air, no wind.

After the null result of the Michelson-Morley experiment, a few physicists still hoped to find some alternative way of measuring the earth's motion through space and the aether. But most scientists began to suspect that there is, even in principle, no possibility of detecting an aether wind. In 1905 Albert Einstein made just such an assumption, and created his enormously powerful special theory of relativity. This theory, which we will discuss in more detail in chapter 9, is based on two assumptions: (1) The speed of light is always the same, and (2) There is no way to detect absolute motion.

The first assumption was, by 1905, pretty well established on the basis of experiments like Michelson and Morley's.

The second assumption is the one that represents a dramatically new way of looking at things. According to Einstein, *there is no possible way in which one can detect motion through empty space.* He jumps here from an experimental fact (that measuring speeds of light rays does not reveal an aether wind) to a blanket hypothesis (that no possible experiment can ever detect an aether wind).

Most of us smile when we hear of the nineteenth-century aether concept. But it is important to realize that insofar as we believe in space as something existing in its own right, we too have an aether concept. Empty space is aether, aether is empty space. But what, then, is the meaning of Einstein's second assumption vis-à-vis aether? Here is what Einstein wrote on this question in his 1920 essay "Ether and Relativity."

> More careful reflection teaches us that the special theory of relativity does not compel us to deny ether. We may assume the existence of an ether; only we must give up ascribing a definite state of motion to it.

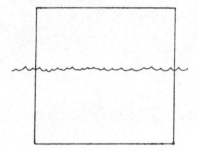

Fig. 73. Space is like a sea's surface.

Think of waves on the surface of water. Here we can describe two entirely different things. Either we may observe how the undulatory surface forming the boundary between water and air alters in the course of time; or else — with the help of small floats, for instance — we can observe how the position of the separate particles of water alters in the course of time. If the existence of such floats for tracking the motion of the particles of a fluid were a fundamental impossibility in physics — if, in fact, nothing else whatever were observable than the shape of the space occupied by the water as it varies in time, we should have no ground for the assumption that water consists of movable particles. But all the same we could characterize it as a medium.

Generalizing we must say this: — There may be supposed to be extended physical objects to which the idea of motion cannot be applied. They may not be thought of as consisting of particles which allow themselves to be separately tracked through time. The special theory of relativity forbids us to assume the ether to consist of particles observable through time, but the hypothesis of ether in itself is not in conflict with the special theory of relativity.

The reader may wonder why Einstein bothers to go through such intellectual contortions. If the aether isn't made up of distinct trackable parts, then why bother with it? Why not just go ahead and say that empty space is pure nothingness? Why pretend that space is some continuous substance called aether? Let's go back to Flatland for an answer.

Unfortunately the passion of the moment predominates, in the Frail Sex, over every other consideration. In their fits of fury they remember no claims and recognize no distinctions. I have actually known a case where a Woman has exterminated her whole household, and half an hour afterwards, when her rage was over and the fragments swept away, has asked what has become of her husband and her children . . .

Even in our best regulated and most approximately Circular families I cannot say that the ideal of family life is so high as with you in Spaceland. There is peace, in so far as the absence of slaughter may be called by that name, but there is necessarily little harmony of tastes or pursuits . . .

EDWIN A. ABBOTT,
Flatland, 1884

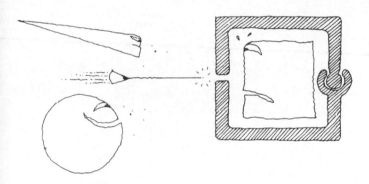

Fig. 74. A Square about to get guillotined.

The Chief Circle decides to let his wife execute A Square . . . She's a bloodthirsty segment, ready and willing to cut poor Square in half. Our hero is shackled in a heavy box with one opening. The Queen surges forward, her sharp end aglitter. She thrusts her point into the box's small opening, thrusts again, and thrusts once more for good measure. But when they open the guillotine box, A Square is as good as new. What happened?

To understand what happened, we should start thinking of the space of Flatland as being somewhat like a rubber sheet or, even better, something like a huge unbreakable soap film. If A Cube seizes a piece of Flatland's space and pulls up, then he can stretch a bit of the space to be bigger than one would expect. And this is just what he did. Cube grabbed the bit of space inside the guillotine box and stretched that space for all he was worth. The Queen's twelve-inch body was not long enough to reach over the "space bump" and get to A Square. Here is how it seemed to Square (I quote again from the imaginary *Further Adventures of A Square*):

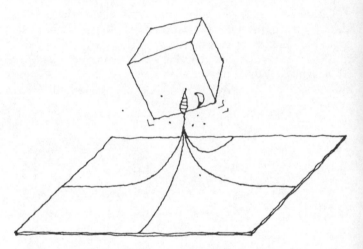

Fig. 75. A Cube stretching Flatland's space.

If my account of what happened is confused, I can only say that this confusion reflects what I and my Countrymen all felt.

The Cube called out to me from Space as the Guillotine Box was fastened around me. Laughingly, he urged me to be composed and of good Cheer. In my unhappy state, this seemed a frivolous, and even an unkind, request.

As the Queen approached, a curious tension thrilled through all my Being. The Box around me seemed to take on more spacious Dimensions. Somehow the hole in the Box's wall grew so deep that the Queen's thirsty point could not attain to my trembling flesh.

Women's sharp Stingers are all but nerveless, and the Queen was not cognizant of her failure. Crying out that the Execution was accomplished, she withdrew. An Isosceles busied himself with the opening of the Box.

But before this was achieved, I was again whirled about my central Axis. My noblest Archetype, the Cube, had now restored me to my original Orientation. As I babbled my thanks, he took yet one more Measure, did one fateful Deed that has ensured my safety from that day on. He reached into the foul Circle's body and crushed the Tyrant's heart.

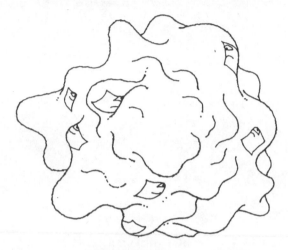

Fig. 76. Lumpy space.

The point of this story is that if we think of space as being made up of a continuous aether jelly, then it becomes meaningful to speak of stretching or distorting space. Although, as Einstein stresses, we must not think of space as made up of particles, it is meaningful to think of space as having bumps and undulations. There is no absolute sense in which one might say that a given bump is moving this way or that, but one can certainly notice how the bumps move relative to each other.

Bumps in space (space*time,* strictly speaking) can be used to explain gravitational attraction. Einstein's 1915 general theory of relativity incorporates a theory of gravity that can best be understood as saying that (1) Matter and energy distort space, and (2) The distortions of space affect the motions of matter and energy. Aether, or space, thus serves as the medium for transmitting gravitational effects. Mass affects space, space affects mass. Let's see how.

We must imagine that the space around any massive body is stretched. The denser the mass, the greater the stretching. A good image for this is of a cue ball resting on a rubber sheet. A sheet sags down around the ball. Or we could think of a helium balloon under the sheet, making it sag up (as drawn in figure 77). The "up" or "down" doesn't matter here; the point is just that the presence of matter stretches space.

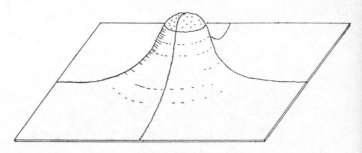

Fig. 77. A massive body makes a bump in space.

Now let's try to see how the curvature of space affects the motion of particles in space. A particularly clear-cut example arises if we let our moving particle be a photon, a tiny piece of light.

Ordinarily we think of light as traveling along straight lines. But if space is curved, there is no such thing as a *really straight* line in space. Nevertheless, light does travel along

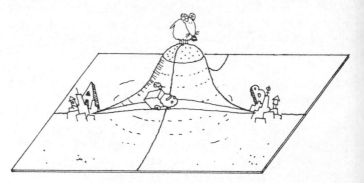

Fig. 78. The shortest path from A to B.

the *straightest possible* lines. Equivalently, we can say that a light ray from point *A* to point *B* will always go along the *shortest possible path* from *A* to *B*.

If there is a big bump in space between *A* and *B*, then the shortest path will not be directly over the bump. The shortest path will be the path that compromises between going right over the bump and looping way out around it. This is easy enough to understand if, in figure 78, we think of *A* and *B* as villages separated by a mountain. The natural, shortest trail between the two is along the wavy line.

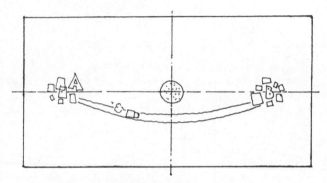

Fig. 79. The shortest path from A to B, seen from above.

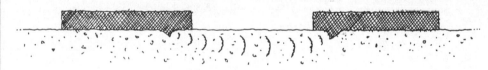

Side view of Astrian telepathy.

PUZZLE 6.1

In Astria, Charles H. Hinton's version of Flatland, it is explicitly assumed that the two-dimensional creatures have a slight thickness, and that they are sliding around on a hard surface. They are like cold cuts lying on a flat tray of space, a space that serves as an elastic medium for transmitting all kinds of vibrations. Deep within the tissue of his or her body, each Astrian has a sharp little 3-D bump, a sort of "astral vibrator." This vibrator, about the size of a phonograph needle, oscillates with the rhythm of each Astrian's thoughts so as to set up sympathetic vibrations in the underlying space. Any other Astrian who is nearby can, without really knowing how, pick up the flavor of these thoughts from the vibrations of his own bump. What kind of impression would the Astrians get of one of their fellows who had been turned over in the third dimension?

Is the physical world in which we live a purely mathematical construct? Put the question in another way: Is spacetime only an arena within which fields and particles move about as "physical" and "foreign" entities? Or is the four-dimensional continuum all there is? Is curved empty geometry a kind of magic building material out of which everything in the physical world is made: (1) a slow curvature in one region of space describes a gravitational field; (2) a rippled geometry with a different type of curvature somewhere else describes an electromagnetic field; (3) a knotted-up region of high curvature describes a concentration of charge and mass-energy that moves like a particle? Are fields and particles foreign entities immersed *in* geometry, or are they nothing *but* geometry?

JOHN A. WHEELER, "Curved Empty Spacetime as the Building Material of the Physical World: An Assessment," 1972

If we look down at this path from above, it looks for all the world as if the gravitational attraction of the massive object had *bent* the light ray by *pulling* on it. Yet what has really happened is that the mass stretched the space so that the shortest path from A to B has a bulge in it. The gravitational bending of the paths of material objects can also be explained in this way, although there the situation gets a little more complicated.

So gravity can be explained by assuming that matter curves space. But why should matter do this? Why should matter curve space?

One explanation is that space curvature is what matter *is*. William K. Clifford first proposed this theory in an 1870 paper called "On the Space Theory of Matter":

I hold in fact

1. That small portions of space *are* in fact of a nature analogous to little hills on a surface which is on the average flat; namely, that the ordinary laws of geometry are not valid in them.
2. That this property of being curved or distorted is continually being passed on from one portion of space to another after the matter of a wave.
3. That this variation of the curvature of space is what really happens in that phenomenon which we call the *motion of matter*, whether ponderable or ethereal.
4. That in the physical world nothing else takes place but this variation, subject (possibly) to the law of continuity.

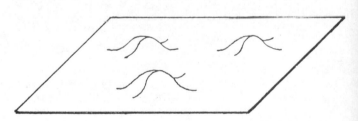

Fig. 80. Three bits of matter in space.

This is a very interesting view of matter, a view that the contemporary physicist John Wheeler has called *geometrodynamics*. Traditionally, people thought of matter as a solid substance floating in empty space. But under the geometrodynamic viewpoint, space is not really empty and matter is not really solid. Space is an aether, a continuous substance that is curved in higher dimensions. And matter is a sort of patterning in the aether.

This idea is intellectually satisfying because it represents the completion of a dialectic triad. Before we had as *thesis* the concept of solid matter, and as *antithesis* the concept of utterly empty space. Matter versus space; something versus nothing. The *synthesis* is to regard space and matter both as a continuous aether substance: when the aether is flat it looks like empty space, when it is sharply curved it looks like matter. The old thesis and antithesis are simply different aspects of the higher synthesis!

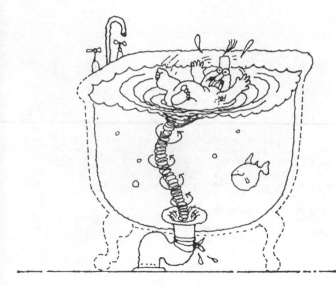

Fig. 81. A vortex thread.

Clifford's notion of building up matter out of pure curved space was a very bold step forward. A few years earlier, William Thomson had taken a partial step in this direction. Rather than taking matter to be higher-dimensional "bumps" in aether or empty space, Thomson proposed that matter is made up of three-dimensional vortex rings in the aether.

A "vortex ring" is something like a smoke ring, a circle of

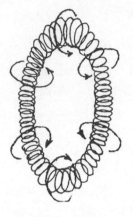

Fig. 82. A vortex ring.

substance that swirls around and around on itself. Thomson's theory was inspired by Hermann von Helmholtz's 1857 proof that in a perfect fluid, any whirlpools or vortices must be centered on lines that either go out to the boundaries of the fluid or curve back on themselves to make circles. In watching the water drain out of a bathtub, one often observes a vortex line of the first type: a wobbly, threadlike whirlpool running from the water's surface to the drain below. The funnel of a tornado is a similar example of a vortex thread. Now if the vortex thread bends back on itself to make a circle, one gets a vortex ring. The remarkable thing about vortex rings is that they consist of a closed-off region of the underlying fluid. This can be seen in watching a smoke ring. For a while at least, a smoke ring does not gain

Fig. 83. A smoke-ring projector. (Engraving from A. E. Dolbear, Matter, Ether, and Motion.)

PUZZLE 6.2

As in puzzle 6.1, we will think about Astrians as being two-dimensional creatures who slide about on top of their space-aether. Each Astrian has a kind of higher-dimensional tooth that digs down into the firm, underlying space. How might the Astrians use meditation techniques to levitate?

or lose any air . . . It consists of the same smoky air circling around and around on itself.

Some nineteenth-century investigators actually built smoke-ring projectors and spent hours watching how smoke rings move, vibrate, and bounce off each other. The hope was that the various properties of matter could all be explained by treating atoms as vortex rings in some perfectly frictionless underlying aether. A particularly nice feature of the vortex-ring theory was that it explained how atoms could have a measurable size, yet be indivisible: a smoke ring has a certain radius, but if you try to cut the ring in half you get nothing but some rapidly dispersed air currents.

The vortex-ring theory of matter failed to lead to any testable predictions, and was eventually abandoned. One of the last books to advocate the theory was a curious work called *The Unseen Universe* (1875). The book seems to say that the *soul* exists as a knotted vortex ring in the aether. By way

If in the body there be no other material than the visible particles, and in the brain no other material than a certain quantity of phosphorous and other things, such as we know them in the common state, and if individual consciousness depends upon the structural presence of these substances in the body and brain, then when this structure falls to pieces there are of course reasonable grounds for supposing that such consciousness has entirely ceased. But it is the object of this volume to exhibit various scientific reasons for believing that there is something beyond that which we call the visible universe; and that individual consciousness is in some mysterious manner related to, or dependent upon, the interaction of the seen and unseen.

BALFOUR STEWART AND PETER GUTHRIE TAIT, *The Unseen Universe*, 1875

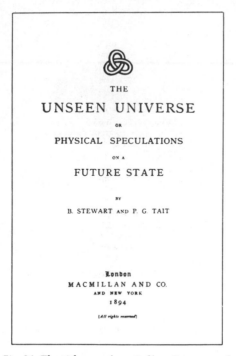

Fig. 84. *The title page from Balfour Stewart and Peter Guthrie Tait's* The Unseen Universe.

Whether this vast homogeneous expanse of isotropic matter [the aether] is fitted not only to be a medium of physical interaction between distant bodies, and to fulfill other physical functions of which, perhaps we have as yet no conception, but also as the authors of *The Unseen Universe* seem to suggest, to constitute the material organism of beings exercising functions of life and mind as high or higher than ours are at present, is a question far transcending the limits of physical speculation.

JAMES CLERK MAXWELL, "The Aether," 1876

of illustrating this notion, the authors (Balfour Stewart and Peter Guthrie Tait) put a picture of a knot on the title page and the binding of their book.

Although it is not immediately evident, the image of a knotted vortex ring involves the fourth dimension. The reason for this is that it is impossible to pass a vortex ring through itself; it has a kind of integrity or solidity akin to that of an actual ring of material substance. Just as Slade and Zöllner claimed that higher-dimensional spirits could knot their sealed cords, the authors of *The Unseen Universe* held that God creates our immortal souls by putting intricate knots and braids into vortex rings of aether. James Clerk Maxwell was amused and fascinated by this notion, and wrote a poem about it:

> My soul is an entangled knot,
> Upon a liquid vortex wrought
> By Intellect in the Unseen residing.
> And thine doth like a convict sit,
> With marlinspike untwisting it,

Fig. 85. "My soul is an entangled knot."

An Astrian slides on space, but a Flatlander is embedded in space.

PUZZLE 6.3

We assume that the Flatlanders are actually in their space, like inkblots in thin paper, or like color swirls in a soap film. A Flatlander cannot exist outside of his or her space. But how, then, was A Cube able to lift A Square out of Flatland's space and turn him over!

Only to find its knottiness abiding;
Since all the tools for its untying
In four-dimensional space are lying.

As the nineteenth century drew to a close, more and more bizarre theories involving matter, aether, and the fourth dimension began to appear — theories that make one think of baroque, overdone ornamentation, of layer after layer of Victorian corsets and petticoats, of opulent, fin de siècle decadence.

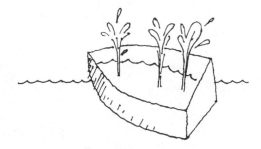

Fig. 86. Water squirts.

In 1892, for instance, Karl Pearson proposed in his *Grammar of Science* that the aether is a four-dimensional fluid seeping into our three-dimensional space — sort of like water welling up through the holes in the bottom of a boat:

In this theory an atom is conceived to be a point at which ether flows in all directions into space; such a point is termed an *ether squirt.* An ether squirt in the ether is thus something like a tap turned on under water, except that the machinery of the tap is dispensed with in the case of the squirt. Two such squirts, if placed in ether, move relatively to each other, exactly like two gravitating particles, the mass of either corresponding to the mean rate at which ether is poured in at the squirt.

PUZZLE 6.4

Suppose we believe, along with Einstein, that it is possible to permanently mark or label a given point in space. But now suppose that we find a hole in space. Doesn't the hole serve to single out a definite space location?

A fourth dimension is involved in Pearson's theory, for this higher dimension is needed to store the "machinery of the taps" that bring the flow of aether to each atom. Two such aether squirts will indeed attract each other because the aether between them is moving more rapidly than in the other parts of space. Let me briefly explain why. There is a law of fluid dynamics called Bernoulli's law, which says that the more rapidly a fluid moves, the lower its pressure is. Since the aether between the squirts moves rapidly, there is a low-pressure region between them, and the rest of the aether will tend to compress this low-pressure region. Thus the two squirts will move together as if drawn by gravitational attraction.

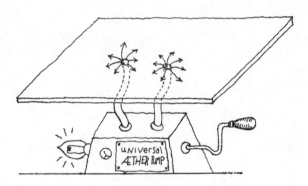

Fig. 87. Two Flatland aether squirts.

PUZZLE 6.5
Quasars are very bright, very distant objects. Recently astronomers found two quasars that seem to be unusually close to each other. Further analysis of these two spots of light revealed that what we are seeing is actually two images of the same quasar. This is explained by assuming there is a massive galaxy between us and the quasar. Can you find a "space bump" diagram that shows how one quasar's image can be split into two?

PUZZLE 6.6
Generally we have represented the space curvature caused by a bit of mass as a rounded-off hump. Suppose that bits of mass were actual points. What kind of space shape would best represent such a mass point?

Crazy theories . . . but none of them crazy enough to be right!

In chapter 11 we'll look at the modern way of building matter up from space-aether. Clifford's idea that matter is a bump in space is basically correct — but he was too conservative in assuming the underlying space to be three- or four-dimensional. According to modern quantum mechanics, a bit of matter is a "bump" in an *infinite*-dimensional Hilbert space!

7

The Shape of Space

I N THE LAST CHAPTER we briefly mentioned the idea that space can be *curved*, curved by bulging it out into the fourth dimension. We looked at two kinds of space curvature: the medium-scale curvature associated with gravitational attraction, and the small-scale curvature that may account for matter. Now we are going to talk about the large-scale curvature of space taken as a whole.

To make quite clear what is meant by this talk of different "scales of curvature," consider the following. On the large scale, we say that the surface of the Earth is curved into the shape of a sphere — a sphere that bulges a bit at the equator. On the medium, human-sized scale, we notice that the Earth's surface is covered with hills and valleys. And on the small scale, the Earth's surface breaks into individual rocks and clods of dirt.

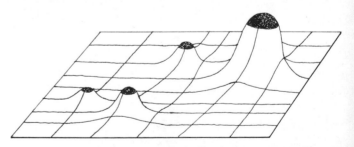

Fig. 88. Medium-scale curvature.

Now, once again, when I speak of the *small-scale curvature of space*, I am thinking of tiny bumps or bubbles or vortices that might conceivably be the same as elementary particles of matter. When I talk about the *medium-scale curvature of space*, I am referring to the planet-sized and galaxy-sized space humps that, according to Einstein, account for the effects of gravitational attraction. And now, when I talk about the *large-scale curvature of space*, I am asking about the overall shape of our universe.

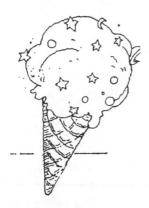

Fig. 89. What is the shape of space?

What is the shape of space? Is it flat, or is it bent? Is it nicely laid out, or is it warped and shrunken? Is it finite, or is it infinite? Which of the following does space resemble more: (a) a sheet of paper, (b) an endless desert, (c) a soap bubble, (d) a doughnut, (e) an Escher drawing, (f) an ice cream cone, (g) the branches of a tree, or (h) a human body?

Questions about the overall shape of space belong to the science called cosmology. I love cosmology: there's something uplifting about viewing the entire universe as a single object with a certain shape. What entity, short of God, could be nobler or worthier of man's attention than the cosmos itself? Forget about interest rates, forget about war and murder, let's talk about *space*.

The ancients seem usually to have thought of our universe as bounded. Either the Earth itself actually had edges, or the Earth was to be thought of as a ball floating inside a large crystal sphere on which the stars were hung. But to a

In cosmology the reliance on physical simplicity, pure thought and revealed knowledge is carried well beyond the fringe because we have so little else to go on. By this desperate course we have arrived at a few simple pictures of what the Universe may be like. The great goal now is to become more familiar with the Universe, to learn whether any of these pictures may be a reasonable approximation, and if so how the approximation may be improved. The great excitement in cosmology is that the prospects for doing this seem to be excellent.

P. J. E. PEEBLES,
Physical Cosmology, 1971

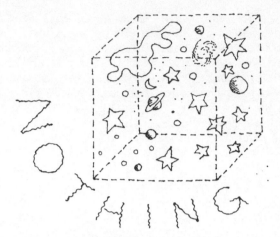

Fig. 90. A finite, bounded universe.

modern thinker, the notion of a universe with edges is almost inconceivable.

What would it be like to come to a place where space stops? Think of a black doorway opening onto Nothingness. Any object that passes through the doorway simply ceases to exist. Beyond the door there is no aether, no space to sustain an object's structure. Such a "door into Nothing" is perhaps a little bit like a star that has collapsed to form a black hole. Maybe there are such doors scattered here and there in our universe. But still, we feel that such doors do not exist on any extremely large scale . . . We do not think that our space, taken as a whole, has edges. In other words, we believe that our space is unbounded.

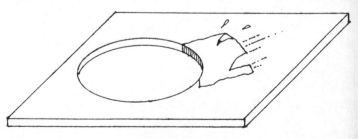

Fig. 91. A hole in space.

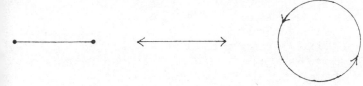

Fig. 92. Three types of 1-D space: finite and bounded, infinite and unbounded, finite and unbounded.

At first blush, one is inclined to think that if space is unbounded, then it must be infinite as well. But this is not the case. In *one* dimension, a circle is an example of a line that is finitely long, yet has no ends. You can walk round a circle forever. The surface of a sphere, such as Earth, is an example of a *two*-dimensional space that is finite and unbounded. In a famous 1854 lecture, "The Hypotheses Which Lie at the Foundations of Geometry," Bernhard Riemann first suggested that something similar is also possible for *three*-dimensional space:

> In the extension of space-construction to the infinitely great, we must distinguish between *unboundedness* and *infinite extent*. The unboundedness of space possesses a greater empirical certainty than any external experience. But its infinite extent by no means follows from this. If we ascribe to space constant curvature, then space must necessarily be finite provided this curvature has ever so small a positive value.

Riemann is suggesting here that our space may be the 3-D hypersurface of a 4-D hypersphere. Back in chapter 3 we talked about what hyperspheres look like from the outside. Now we want to try to imagine what a hypersphere looks like from a point on its hypersurface. We turn, of course, to A Square. What would it be like for the Flatlanders to live not on a plane but on the surface of a 3-D sphere?

The theme of "Sphereland" has been treated many times

Where the sun's rays grazing the earth in January pass off and merge into darkness lies a strange world.

'Tis a vast bubble flown in a substance something like glass, but harder far and untransparent.

And just as a bubble blown by us consists of a distended film, so this bubble, vast beyond comparison, consists of a film distended and coherent.

On its surface in the course of ages has fallen a thin layer of space dust, and so smooth is this surface that the dust slips over it to and fro and forms densities and clusters as its own attractions and movements determine.

The dust is kept on the polished surface by the attraction of the vast film; but, except for that, it moves on it freely in every direction.

And here and there are condensations wherein have fallen together numbers of these floating masses, and where the dust condensing for ages has formed vast disks.

CHARLES H. HINTON,
"A Plane World," 1884

PUZZLE 7.1

Although no line on a curved surface is really straight, some lines are straighter than others: straighter in the sense of being shortest paths. Such straightest possible lines are called the geodesics of the surface. What kinds of lines do you think are geodesics on a sphere?

before. What I'd like to do here is to approach the theme from a new angle.

By way of introduction to the tack I'm going to take, I should confess that this summer (it's October now) I was, for a time, happily obsessed with the fictional notion that Flatland really exists, and that Edwin Abbott used to go look at it all the time. Normally, we think of Flatland as an infinite plane, but so far as we can see, there are no glistening, polygon-filled infinite plans floating around anywhere near the planet Earth. Since we don't see Flatland, I reasoned that it must be hidden somewhere — in a basement, let us say. So now I had an interesting problem to think about: How might one fit an unbounded two-dimensional world inside an ordinary three-dimensional room?

Fig. 93. Let's put Flatland in the basement!

Well, I thought of three ways to fit Flatland in the basement. Here's the first, presented as excerpts from the first-person account of one Arnold Klube. As the first excerpt begins, Klube is about to find Sphereland in the basement of an abandoned technical school.

EXCERPTS FROM ARNOLD KLUBE'S *God of Sphereland*

The door closed behind me and I moved carefully down the dusty stairs. At the bottom I found a light switch. Slowly, hesitantly, I snapped it on.

The small basement room was dominated by a large sphere, apparently weightless. Nearly two meters in diameter, the sphere hovered a few centimeters above the floor. Could this be Flatland?

I moved closer and examined the object's shimmering surface. It was like the taut skin of a large soap bubble: transparent, yet patterned with shifting flecks of color. At first all was confu-

Fig. 94. Sphereland in the basement.

sion, but as I looked closer I began to make out little avenues along which the color flecks moved. These bright, darting animacules were undoubtedly the Flatlanders.

On one side of the room I found a bench lined with scientific instruments, most notably a stand-mounted binocular microscope. Trembling a bit in my excitement, I set the microscope up next to the wonderful sphere.

I will present the facts as succinctly as possible. The world which I have discovered is a two-dimensional film curved into the shape of a sphere some five meters in circumference. The inhabitants of this world — which I call Sphereland — are small polygonal dots, with an average width of one-tenth millimeter. Their space is thus of a circumference equal to some 50,000 body lengths. By way of comparison, note that 50,000 human body lengths comes to 100 kilometers.

Before long, I learned to read the "lips" of the Spherelanders, and to understand their language. As Abbott has reported, they are under the impression that they are living in an infinite plane! It is easy enough for us to imagine walking 100 kilometers, but the fact of the matter is that no Spherelander has ever made the journey "around" space.

There are good reasons for this. If we recall that the surface area of a sphere is given by the formula E^2 / π, where E is the sphere's equatorial circumference, then it is easy enough to cal-

Fig. 95. Arnold Klube.

culate that Sphereland has space to accommodate less than one billion of its citizens, even were they to be packed edge to edge. So far as I can estimate, the actual population of Sphereland numbers some fifty million souls. Thus, each of the Sphere-landers has at his or her disposal an amount of empty space but twenty times that of his or her body — the equivalent, in our terms, of a low-ceilinged prison cell just long enough to lie down in.

Sphereland, in short, is extremely crowded. The entire space is filled with bodies and buildings. The crooked little lanes are as packed with life as any Far Eastern bazaar. Thieves and murderers are everywhere, and to travel any great distance is a virtual impossibility.

Days passed into months, and still I hovered over Sphereland, attentive as an idle god. I found on the workbench certain tools, apparently designed for manipulating the little creatures' space.

Upon experimentation, I learned that the space of Sphereland is not quite two-dimensional after all. It has a definite, though all but imperceptible, thickness. Using some special tweezers and a small cutting frame, I was actually able to take out and examine bits of the space. In one case, I lifted out a certain Square, turned him over, and set him back.

The altered Square's appearance was such that his fellows sought to destroy him. I rescued him and annihilated his chief oppressor. The Square is quite devoted to me now and believes, having seen certain objects which I thrust into his space, that I am a Cube.

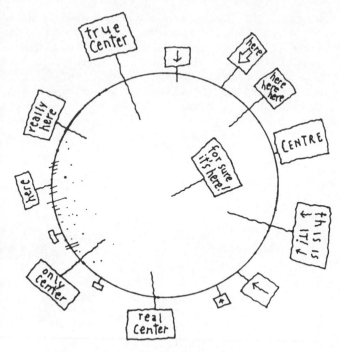

Fig. 96. *Each point on a sphere is equally central.*

Many present-day scientists feel that our space is indeed curved into a hypersphere. Recall from the last chapter that we can take Einstein's general theory of relativity to mean that matter curves space. Apparently, if there is enough matter in our universe, then the cumulative curvature will be enough to bend our space back on itself. If space is hyperspherical, then there are a limited number of galaxies, yet no galaxy is out on the edge of things. Each galaxy is in an equally central position. This is analogous to the fact that any country on Earth can view itself as the central point on the planet.

If our space is hyperspherical, then flying a spaceship in any fixed direction long enough should eventually lead back to our galaxy. Unfortunately, the circumference of our hypersphere is so great that it is unlikely that any Magellan of space will ever carry out such a circumnavigation. One estimate is that our space is about eighty billion light-years in circumference!

Just to get a good image of hyperspherical space, imagine yourself to be floating in a space that's been curved into the shape of a hypersphere a hundred meters in circumference. You're the only object in the space, so it's really dark. You fumble a flare out of your spacesuit, light it . . . and for some reason there's *two* lights. There's the flare right next to you, and there's a flare fifty meters off! What's more, there's somebody else in the space with you, a spacesuited person holding the other flare.

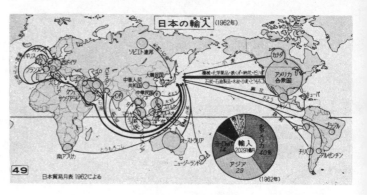

Fig. 97. An un-American world view.

PUZZLE 7.2

Suppose we were to discover a big bright star with nothing inside it but light and empty space. What might we conclude?

PUZZLE 7.3

Most cosmologists assume that any one region of our universe is more or less like any other region. This assumption is known as the cosmological principle. *There is no overwhelming body of evidence for the cosmological principle. People just like it because it makes things simpler. But now suppose the cosmological principle is wrong. Suppose that there is a single most important object in our universe — a unique mammoth object that is very much more massive than anything else. If you combine this supposition with the assumption that space curves back on itself like a hypersphere, what kind of universe do you get? Can you draw a Flatland / Sphereland–style picture of such a space?*

You decide to go visit the other person with the flare. You leave your flare behind, just floating there, and push yourself through the empty space by means of a little hand-held jet you happen to have handy. Your flare and the other person's flare stay immobile . . . but now he's running away from you! He — or is it she? — is upside-down relative to you, and no matter which way you go, he (or she) changes direction to keep you from getting any closer. Could the other person be some sort of mirror image?

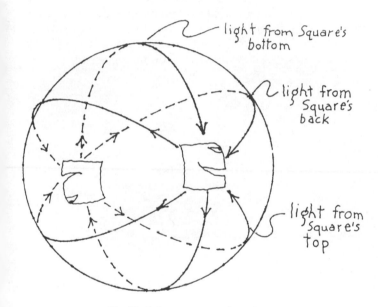

Fig. 98. A Square sees a ghost image.

Yes. Think of A Square on the surface of a small sphere. We assume that the Sphereland light rays move around the spherical space in great circles. What we can notice in figure 98 is that all the light rays that start out from Square's body recross each other at the opposite side of the sphere. This means that Square will see a bunch of images of pieces of his body over on the other side of the sphere, and as it turns out, these images will fit together to make a ghost image of himself, upside-down and mirror-reversed.

To continue. You decide to go look at the other person's flare. It floats there, blazing away, but when you reach out to touch it you find that nothing is there. Why? Because the "other flare" is in fact a virtual image of your real flare. It is a ghost image formed at the place where all the light rays from your real flare cross each other.

Curiouser and curiouser. After returning to your flare you decide that it's getting too cramped in your spacesuit. You take out a tremendous rubber balloon, crawl inside it, and begin filling it up with air from your tank. You've brought the flare along, so the balloon is all lit up inside. It's nice to be inside the balloon and not to have to see the weird ghost images of the flare and yourself. You take off your suit and loll against the balloon's gently curving wall. The tank next to you is hissing away, and the balloon is growing.

Fig. 99. A Square in the weird balloon.

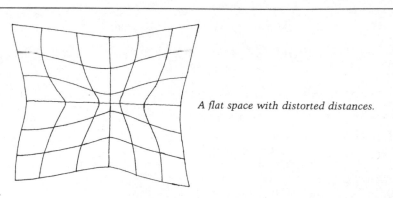

A flat space with distorted distances.

PUZZLE 7.4

Here is a two-dimensional pattern of lines. Suppose that in 3-D space you were to stretch this surface so that the distances between each neighboring pair of lines became the same. What shape would the surface take?

Suddenly, something is wrong. The balloon wall goes flat and starts curving away from you. You are outside the balloon! The tank is still next to you but the air seems thinner. The balloon rapidly shrinks away from you, down to beach-ball size, and collapses. What happened?

Push it down a dimension and look at A Square. A Flatland balloon is just an elastic circle. When it expands past the equator it is able to shrink back down to starting size . . . on the other side of spherical space.

This is a really odd feature of hyperspherical space. Any sphere that expands long enough ends up by expanding past the maximum possible size and continuing to "expand" on down to point size. You can visualize this in our own universe by thinking of a fleet of survey ships flying away from Earth in different directions. If they travel at the same speed, the ships will lie on the surface of an expanding sphere centered on Earth. After a while the pilots will notice, to their surprise, that although no one has changed course the ships are drawing closer and closer together. The unsurveyed space has become a shrinking sphere! When the ships meet up they'll be as far from Earth as it's possible to get. And there will be no surprises left.

So now we have a pretty good image of at least two kinds of space: regular old-fashioned flat space, and finite unbounded hyperspherical space. By "flat" space I mean nothing more complex than a standard three-dimensional Euclidean space stretching out to infinity in every direction. A two-dimensional flat space is called a plane, and a three-

A Möbius strip.

PUZZLE 7.5

A Möbius strip is formed by taking a strip of paper, giving it a half-twist, and taping the two ends together. Think of A Square as an ink pattern that soaks right through the paper of a Möbius strip. If he slides around the strip, what happens to him?

I dozed off and dreamed. But, surprisingly enough, this time I did not see a picture of Lineland — the place which I used to visit as a much wiser Flatlander, bent on telling its citizens the truth, so obvious to myself since I could see the real relations which the people were unable to perceive — no, I dreamed something quite different. I was a Sphere from the country of three dimensions and I was visiting my own world, my own Flatland. No, not Flatland, but Sphereland, because I could see clearly now that my world was curved in a direction which had never been visible to me before . . .

I look to the left, to the right, to all sides, and still the world, my world, does not stretch infinitely in every direction. Of course not, because my world is not infinite. It does not stretch out on all sides into infinity. It is a bend, a curved world. I can go around it! I can go and fly all around my world, my spherical Flatland. How utterly fantastic . . .

dimensional flat space is sometimes called a *homaloidal* space.

What homaloidal space and hyperspherical space have in common is that they are curved the same amount everywhere. Similarly, we say that planes and spheres are *surfaces of constant curvature*. The sphere is certainly not flat, but each point on a sphere is like any other point — there are no lumps. One way of characterizing surfaces of constant curvature is to say that only on such surfaces is it possible to slide triangles around without having the sides and angles change. By the same token, we say that a three-dimensional *space of constant curvature* is a space in which one can move a rigid body around without having the relative proportions of the body change.

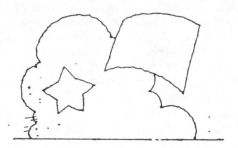

Fig. 100. Gravitational space bending.

The fact of the matter is that our space is *not* a space of constant curvature. A square that is transported to a point near a star will be distorted, by gravitational space bending, into the shape of a curved rectangle. But many cosmologists like to assume that at least on the large scale our space has constant curvature. On the medium scale, Earth's surface has anything but constant curvature; but on the large scale Earth's surface is approximately a smooth sphere — perhaps our space's gravity bumps also average out in a smooth way.

If we do assume that our space has an approximately constant curvature on the large scale, then what are the possibilities? We have mentioned two: flat (or homaloidal) space, and hyperspherical space. There is a third kind of space with constant curvature. This kind of space is usually called *hyperbolic space*. In order to understand what three-dimensional hyperbolic space is like, let's first look into two-dimensional hyperbolic space. A little while back we were thinking about a Flatland that is the surface of a sphere. It turns out that two-dimensional hyperbolic space is best represented as the surface of a so-called *pseudosphere*. But what is a pseudosphere?

A pseudosphere, like a sphere, can be thought of as a 2-D surface. The sphere is smaller than a plane: it bends back on itself and is finite, whereas the plane is, of course, infinite. A pseudosphere, strangely enough, is *bigger* than the plane. Both plane and pseudosphere are infinite, yet the pseudosphere manages to have more room. To begin to get the notion of a pseudosphere, you might think about crawling around on an endless taffy plane. Every few feet you stop to grab and stretch the plane's material. It gets baggy and wrinkled . . . more and more spacious.

Precisely because the pseudosphere is actually bigger than the plane, it is very hard to represent it in the normal Euclidean space of our drawings. But there is a special trick for shrinking a pseudosphere to fit inside a disk. In order to better understand how such a shrinking procedure works, let's consider how one might shrink the normal *plane* to fit inside a *square*.

But how about those light rays? How do they travel? Straight ahead? In straight lines? No, of course not. They cannot leave the space, their space. They have to follow the curve of that space because they belong to it. To us, Spherelanders, they look like straight lines. We think that light travels in straight lines. And they are not really curved in our space, but follow the curve of our space. They have to. But seen from the outside, they are not straight lines, they are nothing but the shortest connecting lines possible on the surface of a sphere.

DIONYS BURGER,
Sphereland, 1965

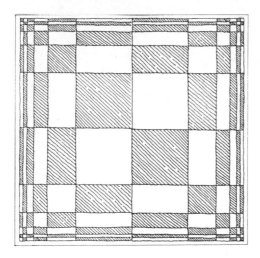

Fig. 101. Infinite checkerboard, shrunk to fit in a square.

I'll present this method as another imaginary first-person account by a man who finds Flatland in a basement. The man is called Felix Ungluck, and, sad to say, he is very clumsy.

EXCERPT FROM FELIX UNGLUCK'S *The Pure Land*

The door swung closed behind me. For a moment I could see nothing, and I worried that I might fall. But then, moving my head back and forth, I caught a gleam of light from below. Something was down there.

I found a light switch and clicked it on. Floating near the foot of the stairs was a weightless square, some two meters on a side. It was marked in an irregular pattern which grew ever more intricate near the fuzzy edges. At first I took it for a sort of flying carpet . . . but then I realized that each colored bit was moving with a life of its own. This was Flatland, its infinite expanse somehow compressed to the size of a two-meter square.

Mesmerized, I descended the staircase and leaned over the shrunken world. The polygonal little Flatlanders darted this way and that, waxing as they neared the center, shrinking as they approached the edges. It was like a living Sunday funnies . . . an *infinite* living Sunday funnies section with smaller and smaller boxes squeezed in on every side, an infinitely regressing pattern in the space of a two-meter square.

Upstairs there was a sudden loud crash. Unsure of my footing, frightened by the noise, I fell forward. My body passed through Flatland as easily as a diver passes through a lake's surface.

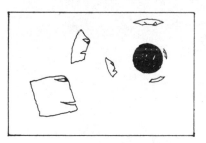

An unreachable hole.

PUZZLE 7.6

Suppose that there is a hole in Flatland, but that no one can fall into it because the closer they get to it, the more they shrink. Can you draw a curved-space picture of this situation?

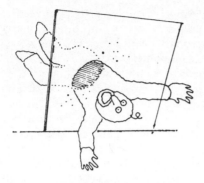

Fig. 102. Felix Ungluck.

The little creatures darted away from my intruding bulk, scattering like so many water striders. And I — like a curse from infinity — I eclipsed an endless acreage and fell heavily to the floor. I seemed to hear the Flatlanders' faint and desperate cries.

Looking down at my body, I realized with horror that I had torn legions of these innocent beings out of their space. They stuck to me, sliding this way and that over the rough fabric of my clothes. I could see that they were mortally wounded. Slowly the pale plaid shimmer died out. I put my face in my hands and began to weep.

Fig. 103. "I put my face in my hands and began to weep."

Figure 101 is not a pseudosphere, but it is based on a trick that will enable us to draw a pseudosphere soon. Figure 101 depicts a plane. The whole infinite plane is squeezed into a finite square. How is this done? By repeatedly halving the distances as one approaches the square's boundary.

Some readers will have heard of Zeno's paradoxes. Zeno was an early Greek philosopher who was interested in demonstrating that the simple idea of "motion" can lead to a number of logical difficulties. His best-known paradox states that you can never leave the room you are in. For, reasons Zeno, in order to reach the door you must first traverse half the distance there. But then you are still in the room, and to reach the door you must traverse half the remaining distance. But then you are still in the room, and to reach the door you must traverse half the remaining distance. But then . . .

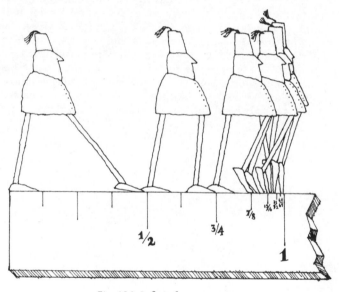

Fig. 104. Infinitely many steps.

The paradox here is that the simple two paces that should get you out of the room can be split into an infinite succession of smaller and smaller paces. In a sense, leaving the room involves completing an infinite task. Of course, we do not normally make each pace half as long as the pace before, and we leave rooms with no difficulty.

But what if you were in a room filled with some strange kind of force field, a force field that made you get half as big each time you took another step toward the door? You would start out two paces from the door, as before. But now after the first pace you shrink to half your size. You are one old pace, or two new paces, from the door. Taking another pace, you shrink again by a factor of two and are now two doubly shrunken paces from the door!

In figures 101 and 105 we show how to use this Zeno-style shrinking to fit an infinite plane inside, respectively, a square and a circle. In figure 101, each of the checkerboard's squares is supposed to be really the same size. Put differently, we can turn the object in figure 101 back into a plane by stretching each of the checkerboard squares back up to unit size. In figure 105, each of the triangles is supposed to be really the same size. One can look at the triangles as being arranged in concentric rings, with each ring half the thickness of the one before. Once again, if we could stretch each triangle back to unit size, we'd get a plane.

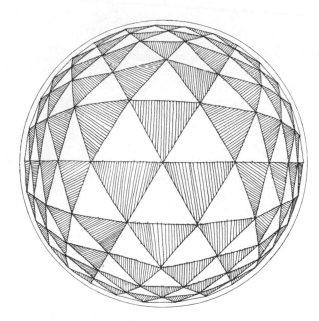

Fig. 105. Infinite plane inside a circle.

Fig. 106. Pseudospherical tesselation.

Now, what about the pseudosphere? The pseudosphere can be represented as an infinite pattern of warped triangles squeezed inside a circle, as in figure 106. The Dutch artist Maurits Escher sometimes used patterns like this. What makes the pseudosphere different is that if you stretched each of its triangles to unit size (and straightened out their sides as well), then you'd get something that will not fit in a plane without folds and wrinkles. Notice, for instance, that at some of the corners in the pseudospherical pattern we have *eight* "equilateral triangles" meeting. But if the pattern could be stretched out and laid flat on the plane, then one could have at most *six* equilateral triangles coming together at a single point.

Although it is impossible to smoothly stretch out the whole pseudosphere in our space, we can stretch out pieces of it to get various bounded surfaces of constant curvature. If we stretch out the central disk of the pseudosphere diagram, it becomes saddle-shaped. Unlike the sphere, which at each point is curving in the same direction, the saddle surface is at each point curving in two different directions. By being thus curved up and down at the same time, the saddle surface stretches to be, in a sense, bigger than a flat disk.

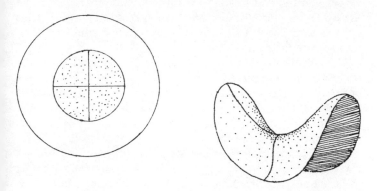

Fig. 107. A disk from the pseudosphere stretches into a saddle shape.

A more dramatic example of stretching out a piece of the pseudosphere occurs if we take a sector that runs out to the "infinitely distant" edge. In figure 108, what we do is cut along lines e, e', and H; then we stretch e and e' out to their "real" infinite length; and then we bend H around into a circle and glue e and e' together to make E. In the 1940s the figure on the right was sometimes called the pseudosphere,

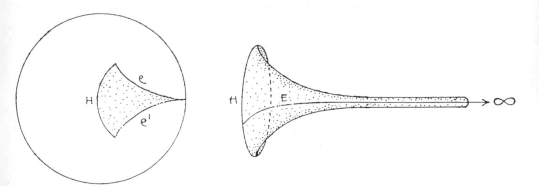

Fig. 108. A sector of the pseudosphere stretches into Gabriel's Horn.

but this is wrong. It is only a sewed-together *sector* of the pseudosphere. What, then, shall we call it? "Gabriel's Horn" seems appropriate, because it is shaped like the bell of a trumpet, and the mouthpiece of the trumpet is infinitely far away. The image I have in mind is of some fearsome Judgment Day rolling around, and an angel in an infinitely far away heaven reaching a long thin trumpet all the way down to blast me in the ear.

So by thinking about what happens if we stretch pieces of the pseudosphere out to proper size, we begin to get an idea of the full surface. Recall now that we started talking about the pseudosphere so as to have an example of a two-dimensional hyperbolic space. But what we really wanted to learn about was *three-dimensional hyperbolic space.*

What is full hyperbolic space like? Imagine an endless three-dimensional space that has somehow been stretched to be more spacious than flat homaloidal space. Mathematically, one can model such a space by taking the inside of a sphere and assuming that the objects inside this sphere shrink endlessly as they move away from the center. This is analogous to our trick of fitting flat or pseudospherical Flat-lands inside circles.

It is a very odd notion to think that perhaps our space is really the incredibly warped and shrunken interior of, say, a tennis ball. As Hamlet says, "O God! I could be bounded in a nutshell, and count myself a king of infinite space, were it not that I have bad dreams."

The idea of representing a curved space by a flat space that is to be stretched and shrunk comes from Bernhard Riemann. As an example of the generality of Riemann's technique, we have indicated in figure 109 how a variety of curved surfaces can be represented by assigning different lengths to the circumference of a fixed unit–radius circle.

PUZZLE 7.7

The Gabriel's Horn surface shown in figure 108 has a very strange property: although its length is infinite, its surface area is finite. Can you think of any way to cut a unit square into infinitely many pieces and arrange these pieces so as to make up an infinitely long surface with an area of one!

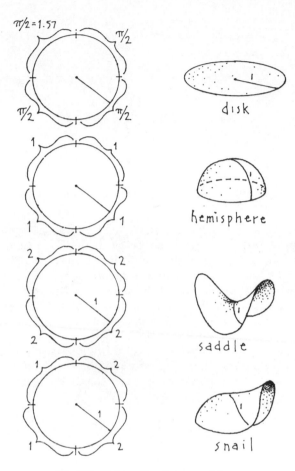

Fig. 109. Making the distances right.

One can make the circumference smaller by bending the surface toward itself, or one can make the circumference larger by bending the surface in two different directions.

If you make a fist you will notice some roughly hemi-spherical bumps where your knuckles are. Pause for a moment and think of the freckles of your skin as being 2-D galaxies in a "Flatland" whose space is the surface of your skin. A galaxy located near your knuckle bumps might have inhabitants who believe that space is spherical. A galaxy in the "saddle" between two knuckle bumps might have citi-

Fig. 110. Your skin is an irregularly curved 2-D space.

zens who feel that space is stretched into a pseudospherical pattern. And the little folks living in the flat expanses of your forearm might take space to be a plane.

In this chapter we have looked at three kinds of 3-D space: flat space, hyperspherical space, and hyperbolic space. What these three kinds of space have in common is that each of them is, on the large scale, uniformly curved. No one region of space is essentially different from any other. But we should keep in mind that the simplifying assumption that our space is of constant curvature might very well be false. The shape of space may be stranger than we thought.

Fig. 111. The shape of space may be stranger than we thought.

8

Magic Doors
to Other Worlds

JUST SOUTH OF BALTIMORE there's a highway exit labeled "Brooklyn." Wouldn't it be great if the exit led right into New York? Even better, wouldn't it be nice to have a special kind of superdoor leading, say, from your living room to the Tuileries gardens in Paris? Or, most exciting of all, how about a hyperdoor leading out of your space and into a totally different universe?

People have always enjoyed thinking about such magical doors. The perfect symbol of the mind's freedom from the body's spatial limitations, magic doors occur throughout fantastic literature, from Lewis Carroll to C. S. Lewis to Robert Heinlein. As a rule, writers of fiction have been very vague about how magic doors might actually be constructed; at best one hears them explained as "tunnels through hyperspace." But as it turns out, modern cosmologists have developed some good ways of thinking about magic doors (also known as Einstein-Rosen bridges or Schwarzschild wormholes).

To get the picture, we turn as usual to A Square. Suppose that Flatland is a plane, and parallel to it is another plane called Globland. Ordinarily, there is no way that an inhabitant of one of these two-dimensional universes can get to the other universe. But suppose that somehow a flap of space from each world has been snipped out, and say that the two flaps are sewn together. Now the Globbers can visit

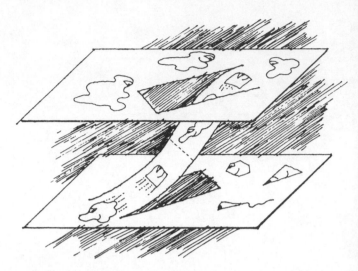

Fig. 112. A space strip connecting Flatland and Globland.

Fig. 113. The Other-Door.

Flatland, and the Flatlanders can visit Globland. I quote once again from that imaginary classic, *The Further Adventures of A Square*:

> The Other-Door which A Cube constructed was most singular. I recognize that to a Spacelander the entire Connection appears to be nothing more than a strip of space stretching from our Land to Globland. But to us the appearance was of the Other-Door as a frameless, unlinteled door giving onto wholly new prospects. This, from the front. From the back, the Other-Door was black Nothingness, a hole in Space. The entire area behind the Other-Door could be approached only at great risk. For here there was no Space at all; from here was taken the Space necessary to build the Way to Globland.
>
> I myself made the journey several times. The Globbers, though Irregular to the last degree, are a pleasant folk, bucolic and accommodating. Several of them ventured into our plane land, although for them the trip was not so easy. Indeed, more than one of them met an untimely end while traveling the Way from Space to Space. Due to a certain awkwardness and grossness of Size, the Globbers found it difficult to avoid brushing against the absolute Nothingness which bounded the Way. And from Nothingness there is no return.

The problem with the world-to-world hookup illustrated in figure 112 is the presence of lethal space edges. But there is a much better way to connect two planes.

After a day's thought, A Cube reappeared in my study. Three highly placed Women were visiting me, and the sudden materialization of the Cube's cross section sent them into an ecstasy of fear. In their eyes I was a magician, the Cube my familiar spirit. I was eager to dazzle these lovely Segments; accordingly I played the wizard's part.

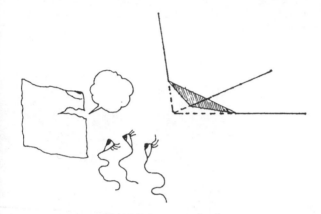

Fig. 114. A Square showing off for the ladies.

I: How now, thou humble Hexahedron?

CUBE: I've thought of a way to fix it, Square. I'll pull the Way around into a tube and get rid of all the edges.

I: 'Tis well. Go then and do my bidding.

CUBE: What kind of talk is that?

I: Begone.

CUBE: Look here, you crummy flat . . .

I: Godspeed, noble Lord.

CUBE: That's more like it. Later.

The apparition vanished, and Una, the loveliest of the three noblewomen, pressed up to me, her natural sway an intoxicating flutter. I promised on the spot to escort her to Globland, or, as she chose to call it, the Astral Plane.

We found the Other-Door quite altered in appearance. Whereas earlier the Other-Door had appeared as a window to Globland from the front, and as a region of Nothingness from behind, it was now the same from every prospect: a lenslike window which seemingly compressed the whole of Globland to the confines of a Disk. My friend the Cube had in some fashion contrived to join together the edges of space where heretofore Nothingness had menaced the world-to-world traveler.

"Oh, Kitty, how nice it would be if we could only get through into Looking-glass House! I'm sure it's got, oh! such beautiful things in it! Let's pretend there's a way of getting through into it, somehow, Kitty. Let's pretend the glass has got all soft like gauze, so that we can get through. Why, it's turning into a sort of mist now, I declare! It'll be easy enough to get through —" She was up on the chimney-piece while she said this, though she hardly knew how she had got there. And certainly the glass *was* beginning to melt away, just like a bright silvery mist.

In another moment Alice was through the glass, and had jumped lightly down into the Looking-glass room. The very first thing she did was to look whether there was a fire in the fireplace, and she was quite pleased to find that there was a real one, blazing away as brightly as the one she had left behind. "So I shall be as warm here as I was in the old room," thought Alice: "warmer, in fact, because there'll be no one here to scold me away from the fire. Oh, what fun it'll be, when they see me through the glass in here and can't get at me!"

LEWIS CARROLL,
Through the Looking-Glass,
1872

Looking into the inside, she saw several coats hanging up — mostly long fur coats. There was nothing Lucy liked so much as the smell and feel of fur. She immediately stepped into the wardrobe and got in among the coats and rubbed her face against them . . .

"This must be a simply enormous wardrobe!" thought Lucy, going still further in and pushing the soft folds of the coats aside to make room for her. Then she noticed that there was something crunching under her feet. "I wonder is that more moth-balls?" she thought, stooping down to feel it with her hands. But instead of feeling the hard, smooth wood of the floor of the wardrobe, she felt something soft and powdery and extremely cold. "This is very queer," she said, and went on a step or two further.

Next moment she found that what was rubbing against her face and hands was no longer soft fur but something hard and rough and even prickly. "Why, it is just like branches of trees!" exclaimed Lucy. And then she saw that there was a light ahead of her; not a few inches away where the back of the wardrobe ought to have been, but a long way off. Something cold and soft was falling on her. A moment later she found that she was standing in the middle of a wood at night-time with snow under her feet and snowflakes falling through the air.

C. S. LEWIS,
The Lion, the Witch, and the Wardrobe, 1960

No one had yet ventured through the altered Other-Door. Desirous of assuring my conquest of Una, I courageously pressed forward to the mysterious Disk. It gave the strange appearance of a circular Mirror, such as ornament our Trees at festival time. Peering into it I could make out smaller and smaller Globbers, ever-dwindling toward the inconceivably distant central Point. I feared to enter, feared that I might be crushed by Shrinkage. But Una was vibrating at my side, urging me on with her low, melodious voice. "Come, Una," I said, and slid forward into the weird Disk which somehow contained all of Globland.

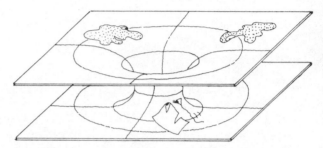

Fig. 115. A Square and Una at the space tunnel's mouth.

The Globbers had appeared quite shrunken and distorted before we entered the Door. But now, as we pressed forward, they took on their familiar, albeit Irregular, appearance. Could it be that we had shrunken to their size? All around us lay the endless expanse of Globland. Was this really the interior of some magical Disk? My thoughts were interrupted by Una's excited cries.

UNA: Oh, look, dear Square, Flatland is now a Disk itself!

I *(turning to look back):* Indeed. Perfect Symmetry prevails. As seen through the Space Tunnel, Flatland is a Disk in Globland, and Globland is a Disk in Flatland. All this have I wrought for your pleasure, my Lady.

UNA: It is well, my Lord. My Husband cannot disturb us here in this enchanted Land.

I: Then let us dally, fairest Una.

UNA: Freely, my Lord . . . yet look into the Disk of Flatland. There slides my Mate, A Hexagon!

I: He is small and puny. He is an ant.

UNA: But, oh, dear Square, he waxes as he nears the Disk's edge!

I *(to a nearby Globber):* How now, sirrah, wilst grant me a boon?

GLOBBER: Blub, yubba, gloop.

I *(seizing him with my Mouth):* Just stretch yourself (mmpf) like this (mmpfmmp) and this, dear friend. And in such wise do curtain our Seraglio.

The deed was done, and Una and I were free to take our pleasure. My perfect content was marred only by one question: How was it that passing through the Space Tunnel turned Inside to Outside, and Outside to In?

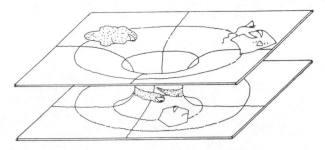

Fig. 116. A Globber blocks A Hexagon's view of Una's trans-dimensional tryst.

Looking at the space tunnel from outside the planes of Flatland and Globland, we can see the answer to A Square's question. The throat of the "wormhole," or space tunnel, is bounded by a circle in either world. A Flatlander looking at this circle sees light from every part of Globland . . . Thus it

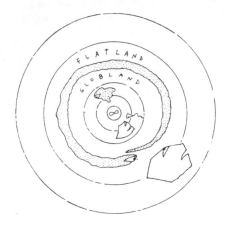

Fig. 117. A Hexagon thinks Globland is a disk.

Joe noticed something shiny lying on the next mound over. A bright little sphere, like a big ball-bearing or a silvered glass Christmas-tree ball. An odd thing to find in an asparagus field.

He hopped over the intervening mound and leaned over the little mirror-ball. The sky was in there, and his face and the horizon and the field. Neat. But . . .

Wait. It wasn't the *same*. The field in the little reflected image was pink and crowded with towering . . . machinery, tapering in towards the image's center. Worse, the funhouse face looking back at Joe was not his after all . . . was not any living human's . . .

The mouth was moving. Calling others. More faces crowded up. Two, three, five . . . small and distorted in the mirror's curve.

Joe gasped and stepped back, then stepped forward and gave the ball a poke. It rolled off the mound. Nothing in the image changed. The central figure was holding up a three-fingered hand and making signs. The vaguely female mouth-slash moved soundlessly. Over the figure's head Joe could make out a tiny rocket-plane moving across the curved sky, moving away and away, dwindling towards the infinitely distant central point. It was a whole universe in there.

RUDY RUCKER,
"The Last Einstein-Rosen Bridge," 1983

seems to him that Globland is somehow compressed to fit inside a circle. By the same token, a Globlander will see light from all over Flatland as coming from the circular throat of the wormhole.

Fig. 118. Hyperspace tunnel to another universe.

Now, as we have done so many times before, let's imagine raising everything by one dimension. Imagine that there is another 3-D universe, "parallel" to ours in 4-D space. If we could move *ana* through hyperspace we could get to the other universe. But we find it very hard to move in the fourth dimension. How, then, could we ever get to the other universe? By traveling through a hyperspace tunnel, a so-called Einstein-Rosen bridge. What would such a hyperspace tunnel look like? The entrance to it would look like a sphere that contained a whole other universe, incredibly shrunken and distorted. If you dived headfirst into this sphere, you would feel as if you passed right through it. But then when you looked around you would realize that you were in the other universe now, and looking back at the hyperspace tunnel, you would see a sphere that seemed to contain your whole original universe, incredibly shrunken and distorted.

There is actually a very familiar object that looks just like the mouth of an Einstein-Rosen bridge: a glass Christmas tree ornament. Such a spherical mirror reflects, in principle, the entire universe around it. The farther an object is from the mirror's surface, the closer its image seems to lie to the mirror's center. Of course, if you were looking at the mouth of a hyperspace tunnel to *another* universe, you would not see the mirror image of our universe, you would see what looks like the mirror image of another universe.

Is there any way — short of the miraculous intervention of a higher-dimensional being — that an Einstein-Rosen bridge could actually come into existence in our cosmos? Yes. If there are indeed some other 3-D universes parallel to ours, it might be that a sufficiently dense object could bulge our space out enough to touch another space. And the two spaces might join together like two soap films that have been brought into contact.

In order to easily illustrate this, let's just work with a cross section of two parallel Flatlands. As has been discussed, the presence of matter causes space to bulge out. Now, just as a woman's high heel will dent a rubber mat more than a man's larger heel, it turns out that *the denser the matter, the greater the space distortion.* If our sun could be compressed to a much smaller size, then it would distort space much more.

The sun is basically a ball of hot gas. The gravitational attraction of the sun's particles for one another works to try to make the sun smaller. The thermal agitation of the hot gas particles works to try to make the sun bigger. The two forces balance out with the sun just the size it is. Eventually, however, the sun will grow cooler. As it cools there will be less outward pressure, and gravity can work to make the sun get smaller and denser. This compression heats the sun back up for a while, but eventually it cools again, and contracts even more.

All stars go through this stepwise contraction process as they age. Depending on the star's starting mass, various final outcomes are possible. If at some point a star contracts too rapidly, it explodes and makes a nova or a supernova. If the star is not too massive to start with, it may contract down to form a solid, glowing lump of metal. If it is a bit more massive, it squeezes down further by collapsing the metal atoms. Protons combine with electrons to give neutrons, and one gets a fantastically dense "neutron star." These stars are made up of a substance called neutronium, which masses about one billion kilograms per cubic centimeter.

The contraction is most dramatic if the star's mass is so

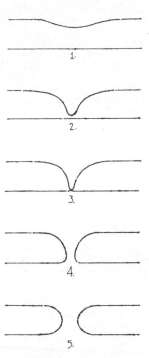

Fig. 119. A collapsing star could form an Einstein-Rosen bridge.

Of all the conceptions of the human mind from unicorns to gargoyles to the hydrogen bomb perhaps the most fantastic is the black hole: a hole in space with a definite edge over which anything can fall and nothing can escape; a hole with a gravitational field so strong that even light is caught and held in its grip; a hole that curves space and warps time. Like the unicorn and the gargoyle, the black hole seems much more at home in science-fiction or in ancient myth than in the real universe. Nevertheless, the laws of modern physics virtually demand that black holes exist. In our galaxy alone there may be millions of them.

The search for black holes has become a major astronomical enterprise over the past decade. It has yielded dozens of candidates scattered over the sky. At first the task of proving conclusively that any of them is truly a black hole seemed virtually impossible. In the past two years, however, an impressive amount of circumstantial evidence in the constellation Cygnus designated Cygnus X-1. The evidence makes me and most other astronomers who have studied it about 90 percent certain that in the center of Cygnus X-1 there is indeed a black hole.

KIP S. THORNE,
"The Search for Black Holes,"
1974

great that even the neutronium ends up getting crushed. In such cases the star shrinks down to a tinier and tinier volume . . . possibly even down to point size! Such superdense collapsed stars are the "black holes" that one so often hears about. The reason for the name is that if a star is dense enough, then its gravitational attraction becomes so powerful that light cannot escape the star. In other words, a star that becomes dense enough takes on the appearance of a black, lightless sphere in space — a region that emits no light at all. Obviously it is difficult to "see" a black hole, but a variety of indirect observations seem to indicate that there really are quite a few black holes floating around in space.

Fig. 120. A black hole absorbs light.

As indicated in figure 119, if a massive star or black hole distorts space enough, it is possible that an Einstein-Rosen bridge to another universe could be created. Flying into the right kind of black hole might pop you out into a different world. The theme of black-hole-as-gateway-to-other-realities is amusingly used in the Walt Disney movie *The Black Hole*. At the end of the movie, the good guys and the bad guys all fall into a huge black hole. The hole turns out to be an Einstein-Rosen bridge with two exits: heaven and hell! This type of idea harks right back to the hyperspace theologians of Abbott's time.

It is also possible to have a wormhole, or E-R bridge, which leads back to the same space it starts from. This could be very important. Here's why.

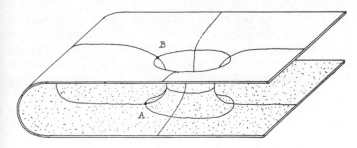

Fig. 121. The wormhole is a short cut from A to B.

According to Einstein's special theory of relativity, nothing can travel faster than light. This has always been a severe limitation for conscientious science fiction writers. It takes light four years to get to Alpha Centauri, our sun's closest neighbor . . . And any conversation or cultural exchange that is riddled with four-year gaps is going to be pretty dull to read about. If you're interested in traveling to another galaxy, the situation is much worse: our nearest galactic neighbor, the Great Magellanic Cloud, is well over ten thousand light-years away!

SF writers have often avoided this problem by supposing that (1) our space is folded back on itself, and (2) there are E-R bridges or wormholes connecting the various folds. By finding the right hyperspace tunnel, a very long journey can be cut down to manageable size. One of the first writers to use this device was Robert Heinlein, the father of modern SF. The analogy is to an ant on a silk scarf. Normally it takes the ant a long time to crawl from one corner to the other, but if the scarf is all crumpled, then by leaving the material's surface the ant can find a short and direct path through 3-D space.

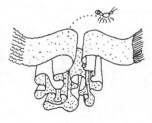

Fig. 122. A hyperspace hop.

Of course, being able to take a desired short cut depends on there being a usable E-R bridge at the right location. Some SF writers avoid this problem by having their space travelers *create* E-R bridges as necessary. In Piers Anthony's mind-boggling *Macroscope*, the method of travel is to get inside a large object (say the planet Neptune), use some miraculous ray to make the object collapse to black-hole size, and then zip through the black hole to come out somewhere different!

If we could really manipulate the curvature of space at

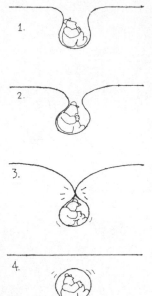

Fig. 123. A piece of space pinching off.

will, then there would be an interesting alternate way of traveling through hyperspace. Instead of building a tunnel or wormhole leading to another space, one could pinch a small hypersphere off our space and just float away. This could be risky, of course, as you'd have no way of predicting where and when your little space bubble might meet another universe. But one nice thing about leaving space by pinching off a closed piece of it is that this doesn't leave a hole behind.

So far in this chapter we have discussed how one might travel through hyperspace to other universes, and we have mentioned how this type of travel might also be useful for finding short cuts from one region of our space to another. One question we have not touched on yet is whether or not there really *are* any other universes.

As was hinted at in the last chapter, it is abstractly possible to treat the gravitational distortion of space as a type of stretching and shrinking of flat space — as opposed to the bulging out of flat space into some higher dimensions. Many scientists feel that "curved space" is just a colorful phrase, and that there really is nothing outside our three space dimensions. For these somewhat cautious thinkers, the visible universe is all that exists, and any talk about alternate universes is just empty dreaming.

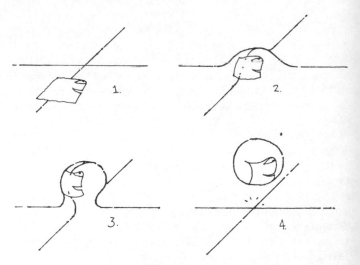

Fig. 124. A Square traveling on a space sphere.

But if we take the fourth dimension quite seriously, then it seems natural to suppose that there might be other universes. All these universes taken together make up a much grander entity, variously known as the cosmos or as superspace. In traditional Christian doctrine, the cosmos has three parallel layers: heaven, our world, and hell. The theosophists hold that the cosmos has seven layers, six of

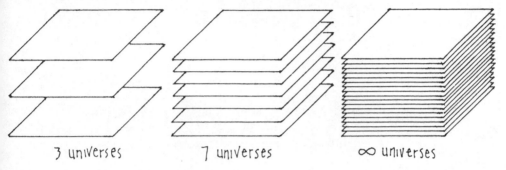

3 universes   7 universes   ∞ universes

Fig. 125. Parallel worlds.

these being "astral." A common notion in science fiction is that there are endlessly many parallel universes — with each possible universe existing somewhere. A variation of this last idea has actually been incorporated into modern quantum mechanics, and we will return to it in later chapters.

Probably the least interesting viewpoint on the question of how many universes there are is that which says: "The whole question is meaningless. No one has any idea how to detect another universe. Since statements about other universes cannot be subjected to immediate scientific testing, these statements don't really say anything at all."

Such a viewpoint combines two assumptions: (1) *Seeing is believing*, that is, if something is real we can find a way to observe it; and (2) *There's nothing new under the sun*, that is, we've already observed every type of thing we ever will observe. The first assumption is central to the philosophical school of logical positivism, a modern outgrowth of traditional British empiricism. For the positivist or empiricist, the world is basically equivalent to the sum total of all possible sensory experience. I have no problem with this —

indeed, I will advocate a similar position in part III. It is the second assumption I object to. No one has *yet* found a way to observe the other universes, granted. But this does not automatically prove that no one will *ever* find a way to "see" the other worlds.

People speculated about atoms centuries before there was any hope of detecting an individual atom. And if no one had ever talked about atoms, the means to detect them might never have been developed. Talking about other universes would be a more respectable pastime if we could already see them. But unless we go ahead and try to imagine ways in which this might happen, the day will never come. We have already seen that something like Einstein-Rosen bridges may exist as actual pathways to other universes. What I want to do now is to think of some other ways in which these universes might make themselves known.

Just so as not to be lost in a sea of possibilities, let's limit ourselves to one fairly reasonable model for the cosmos: four-dimensional hyperspace with a number of hyper-spheres floating in it. Each of the hyperspheres makes up a single universe. We might think of these hyperspheres as being something like bubbles in a fluid; alternatively, we might think of them as being like planets floating in space.

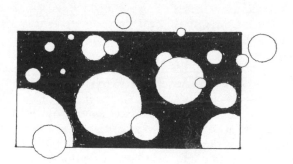

Fig. 126. Hyperspherical universes floating in hyperspace.

PUZZLE 8.1
What would happen if the Globber in figure 116 were to choke the throat of the space tunnel down to point size?

Fig. 127. Two spaces merging.

In terms of two-dimensional beings, we are thinking of a bunch of Spherelands.

If we are limited to the 3-D hypersurface of our hypersphere, is there any way in which we can become aware of the other hyperspheres? One very dramatic way in which this could happen would be if one of the other universes happened to bump into ours. Imagine what a Spherelander would see if his space bubble were suddenly to bump into and join up with another space bubble. The effect would be, in our terms, as if all the visible stars were to move out toward the horizon, leaving room for a whole lot of new stars at the zenith! Of course, if the other space was considerably smaller than ours, the effect would be less dramatic. If a small hypersphere merges with ours we might perceive this joining-up process as the occurrence of an unusually bright spot in the sky. Conceivably, the very distant and very bright light sources known as quasars (for "quasi-stellar objects") are spots where small energy-filled hyperspheres are in the process of joining up with our own hypersphere.

But are there any less large-scale and less obvious effects that might point to the existence of other hyperspherical spaces? In thinking about this question, it is helpful to imagine the situation of a man blind from birth. Suppose that he takes it into his head that the sun, the moon, the other

Fig. 128. Quasars!

PUZZLE 8.2

An Einstein-Rosen bridge would look something like a spherical mirror, with the odd property that the world in the mirror was actually different from the world outside the mirror. Now imagine an ordinary flat mirror with the property that the world seen in the mirror is not the same as our world on our side of the mirror. What kind of connection between two spaces is being described here?

Fig. 129. Blind and stubborn.

planets, the stars, and so on do not exist. Suppose he insists that space is a vast emptiness containing but one object: the spherical planet Earth. How might you convince him he is wrong?

Offhand I can think of three approaches: (1) You might teach him to be sensitive enough to heat radiation so as to "feel" the sun's passage across the sky. Or you might couple a telescope to a photocell that controls the volume of a little buzzer. Moving this telescope around, the man would learn to perceive the stars as "loud" spots. (2) You could get him to notice the rise and fall of the tides, and explain to him that this is caused by the gravitational pull of the moon. (3) You could get him to notice the various effects of the Earth's rotation: the equatorial bulge, the so-called Coriolis force, and the existence of poles. And then you could argue that if the Earth is really rotating, it must rotate relative to some other celestial objects.

Let's consider the higher-dimensional analogues of these three sorts of ways to notice other worlds.

1. Unless another hypersphere actually touches ours, there is no way for light to leave its space and come to ours. So we cannot hope to see it. So far as we know, any other kind of radiation would also be confined to the 3-D space where it originates. So it does not seem very likely that any kind of training or equipment can help us "see" the other hyperspheres. Our position is, after all, really like that of a polygon on a Sphereland . . . not like that of a blind man on Earth. And there seems to be no reason that any one Sphereland would send radiation to other Spherelands. A second difficulty here is that even if some higher-dimensional radiation from other spaces did fall on our space, the radiation would not be focused at any particular spot. At best, those regions of space that lie nearer to the other hyperspheres would be observed to have more radiation in them.

2. Gravity is not so much a type of radiation as a condition of space. It is conceivable that there could be a higher-dimensional analogue of gravitation, according to which various four-dimensional objects would distort the hyperspace they float in. Just as the moon's motion around the Earth causes a bulge to travel around the Earth's surface, we can suppose that a nearby hypersphere might distort the shape of our own hypersphere. But present-day scientific apparatus is not even sensitive enough to measure our hy-

persphere's radius — let alone "tidal" variations of the same radius.

3. This approach is probably the most important one. Given that nonrotating spheres are virtually unheard of in our universe, it seems rather likely that the hypersphere that makes up our space is itself rotating. And, as I will now argue, if our space is rotating, then it is almost certain that there are spaces outside our own. This last move will not seem immediately obvious: it is based on a fairly unfamiliar notion called Mach's principle. Ernst Mach (1838–1916) formulated this principle to account for objects' *inertia*, their tendency to resist being moved. The point Mach makes is that if an object were totally alone in empty space, then it would be meaningless to say the object is rotating or being accelerated. An object alone in empty space would, in effect, have no inertia, no heft, no resistance to being moved. Therefore, argues Mach, the fact that an object on Earth has heft is a consequence of the existence of all the distant stars and galaxies. By the same token, the fact that we notice the Earth's rotation is also the result of the existence of the distant stars. Generalizing Mach's principle to hyperspace, we conclude that if we can find evidence that our hyperspherical universe is rotating, then we have good reason to believe that there are other universes, *relative to which* we are rotating. O.K. Now the question is this: What kind of evidence for the rotation of our universe might we hope to find?

Fig. 130. *Spinning globe with hurricanes.*

"I don't know exactly how to tell you; it takes equations. Say! Could you lend me that scarf you're wearing for a minute?"

"Huh? Why, sure." She removed it from her neck.

It was a photoprint showing a stylized picture of the solar system, a souvenir of Solar Union Day. In the middle of the square of cloth was the conventional sunburst surrounded by circles representing orbits of solar planets, with a few comets thrown in. The scale was badly distorted and it was useless as a structural picture of the home system, but it sufficed. Max took it and said, "Here's Mars."

Eldreth said, "You read it. That's cheating."

"Hush a moment. Here's Jupiter. To go from Mars to Jupiter you have to go from here to here, don't you?"

"Obviously."

Well . . . How can the blind man tell that Earth is rotating? A major effect of the Earth's rotation is that the Earth, instead of being a perfect sphere, bulges at the equator and is flattened at the poles. But, as was already mentioned, we can't even measure our universe's curvature yet — let alone the deviation of this curvature from perfect hypersphericity. Another very important effect of the Earth's rotation is the existence of the Coriolis force. It is the Coriolis force that causes hurricanes to rotate clockwise in the Northern Hemisphere and counterclockwise in the Southern Hemisphere.

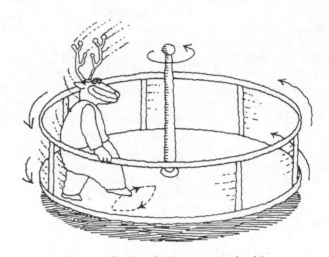

Fig. 131. The Coriolis force in everyday life.

How does the Coriolis force work? To see it in operation, go to a playground that has one of those hand-pushed merry-go-rounds. Get on, crank it up, face the center, and let your right leg swing back and forth. What you will notice is that

PUZZLE 8.3

In the last chapter, I said there were three ways to fit Flatland into a basement, but I only described two ways: bend Flatland into a sphere, or infinitely shrink it to fit in a square. What is the third way?

when your leg moves toward the center, some mysterious force pushes it to the right, and when your leg moves away from the center, a force pushes it to the left. This is a Coriolis force, resulting from the merry-go-round's rotation relative to the rest of the universe. If the merry-go-round were the only object in the universe it would be meaningless to say it is rotating and you would, in fact, feel no Coriolis force. Returning to the swinging leg for just a moment, note that the net result of the Coriolis force is to turn a simple swinging motion into a clockwise circling motion. It is the same force, on a larger scale, that causes air masses to spiral in a clockwise fashion in Earth's Northern Hemisphere.

The fact that hurricanes tend to turn different ways in different parts of the world is a definite asymmetry. If we could find some pervasive asymmetry in the behavior of matter throughout space, we might suspect that this asymmetry results from some overall property of the universe. I have no very specific suggestions as to what kinds of asymmetries we should look for, but one general principle is that we should expect to find the crucial asymmetry in the domain of very small particles. Why? My feeling is that if our space has a slight thickness in the fourth dimension, then very small particles may enjoy a certain four-dimensional freedom of motion, and thus be more susceptible to influences from hyperspace. The situation is complicated by the fact that the rotation of a hypersphere in hyperspace is more complex than a sphere's rotation in space.

But enough of these subtleties. Whether there really are other universes is not nearly so important as the realization that we are integral parts of this one. The fabric of space joins us all together; we are all ripples on the bosom of the aether sea. Space is not some dead abstraction, but rather a living and moving thing. We are patterns in space, and, strangely enough, it is not beyond our abilities to speculate on the overall shape of space itself.

"But suppose I fold it so that Mars is on top of Jupiter? What's to prevent just stepping across?"

"Nothing, I guess. Except that what works for that scarf wouldn't work very well in practice. Would it?"

"No, not that near to a star. But it works fine after you back away from a star quite a distance. You see, that's just what an anomaly is, a place where space is folded back on itself, turning a long distance into no distance at all . . . The math of it is simple, but it's hard to talk about because you can't see it. Space — *our* space — may be crumpled up small enough to stuff into a coffee cup, all hundreds of thousands of light-years of it. A four-dimensional coffee cup, of course."

ROBERT A. HEINLEIN,
Starman Jones, 1953

Part III
HOW TO GET THERE

9

Spacetime Diary

Monday, November 15, 1982

I F IT WASN'T for time, I could live forever. Does that make sense? If it wasn't for space, I could be everywhere. Is there a difference? I want to go back to my happy college days. I want to be a newlywed again. I want to be three feet tall and sit on my mother's lap. I don't want to die. I want to see the future. Time won't let me. Let's kill time. Let's get past time. Let's reach through time and grab hold of eternity. Now there's no time. There's no time now.

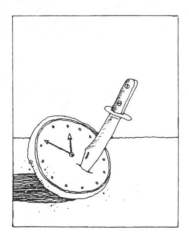

Fig. 132. Let's kill time.

Later. Do you hate time? Alarm clocks, sure. Changing the clocks for daylight-saving time is the worst. How can they just take an hour away like that? Remember in 1978 when Nixon took away *two* hours for the oil companies?

"The older I get, the faster time goes," my mother told me. "The years just fly by. Every time I turn around, it's Christmas or Thanksgiving." Party time speeds up and slows down like an out-of-control movie. Ten minutes lasts for two hours, but the next time you look at your watch, it's three in the morning. Airport time. Sex time. Street time. Fast or slow, it all passes.

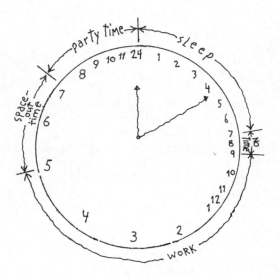

Fig. 133. Fast and slow.

That was my big realization twenty years ago. It all passes. Here I am at the bathroom door, and how can I ever get to the sink? How can high school ever end, how can I ever finish college, how can I ever be married? But then I'm at the sink, I'm back out the door, I have a Ph.D., I'm married with three kids, and twenty years have passed. Here I am alive, and how can I ever die? But I will, I know I will, I know it in my soul.

Death. It's like the basic puzzle issued to each of us at birth. *Hi, you're alive now, isn't it nice? Someday you'll die and it'll be over. What are you going to do about it?* It's

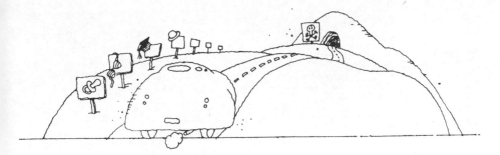

Fig. 134. It all passes.

awful, it's terrifying, it's enough to make a person commit suicide!

If time didn't pass, I'd always be here now, writing this chapter. I'm scared of dying. I'd like to think that time doesn't really pass. What I'm going to do in this chapter is to present some scientific justifications for the belief that the passage of time is an illusion.

People ordinarily think of the world as being a three-dimensional space that changes with the passage of time. The past is gone, the future doesn't exist yet, and only the present is real. But there is another way of looking at the world: we can regard the world as a *block universe*. When we think of the world as a block universe, we put all of space and time together to make a single huge object. The block universe is made up of *spacetime*. Spacetime is four-dimensional: three space dimensions plus one time dimension. To look at spacetime from the outside is to stand outside of history and view things *sub specie aeternitatis*.

"Spacetime" may sound like something technical and far removed from ordinary life. But I would argue that it is really a more natural concept than "space that changes with time."

Suppose you work in an office miles away from your house. At 7:00 you see your bedroom; at 10:00 you see your desk. One day at 10:00 you sit in your office and wonder what is real. If you believe the world consists of a space that changes with time, then you are more or less committed to the view that the past is gone. So you will feel that your 10:00 bedroom exists, but your 7:00 bedroom does not exist. Yet your 10:00 bedroom is not something you can *see*, sitting there in your office. Wouldn't it be more reasonable to

Philosophia perennis — the phrase was coined by Leibniz; but the thing — the metaphysic that recognizes a divine Reality substantial to the world of things and lives and minds; the psychology that finds in the soul something similar to, or even identical with, divine Reality; the ethic that places man's final end in the knowledge of the imminent and transcendent Ground of all being — the thing is immemorial and universal.

ALDOUS HUXLEY,
The Perennial Philosophy, 1944

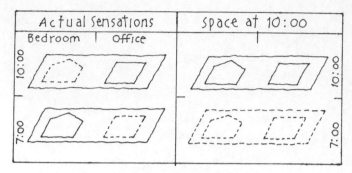

Fig. 135. What's real?

believe that the 7:00 bedroom (which you *saw* and can well remember) is real, and that it is the 10:00 bedroom whose existence is doubtful?

My world is, in the last analysis, the sum total of my sensations. These sensations can be most naturally arranged as a pattern in four-dimensional spacetime. My life is a sort of four-dimensional worm embedded in a block universe. To complain that my lifeworm is only (let us say) seventy-two years long is perhaps as foolish as it would be to complain that my body is only six feet long. Eternity is right outside of spacetime. Eternity is right now.

This is not a new idea by any means. The teaching that all history is an eternal Now is central to the classic mystic tradition. In one of his sermons, the fourteenth-century priest Meister Eckhart expressed the basic idea as vividly as anyone before or since:

> A day, whether six or seven ago, or more than six thousand years ago, is just as near to the present as yesterday. Why? Because all time is contained in the present Now-moment.
>
> To talk about the world as being made by God tomorrow, or yesterday, would be talking nonsense. God makes the world and all things in this present now. Time gone a thousand years ago is now as present and as near to God as this very instant.

Whether or not we share Eckhart's beliefs about God, the image of spacetime being created all at once is a powerful one. Whenever I read his words I get an image of a big old man with a white beard flinging a bucket of paint at a barn wall. *Splat*: there's all of spacetime, created all at once, created right now.

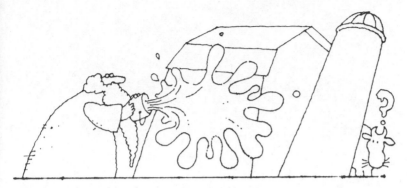

Fig. 136. "God makes the world and all the things in this present now."

Tuesday, November 16, 1982

Space is made up of *locations*; spacetime is made up of *events*. An "event" is just what it sounds like: a given place at a given time. Each of your sense impressions is a little event. The events you experience fall into a natural four-dimensional order: north/south, east/west, up/down, sooner/later. When you look back at your life, you are really looking at a four-dimensional spacetime pattern. So there is nothing very strange or confusing about spacetime, as long as we are looking at it from the "inside."

Looking at spacetime from the "outside" is a little harder: four-dimensional things are always difficult to visualize. Let us, once again, think about Flatland. Imagine that A Square is resting alone in an empty field, and that shortly after noon his father, A Triangle, slides up to him and then slides off. If we take time to be a third dimension perpendicular to the plane of Flatland, then we can illustrate these events by a spacetime diagram as shown in figure 137. Here A Square and A Triangle are wormlike patterns in spacetime. Their brief encounter at 12:05 is represented as a bending together of their lifeworms. Nothing really moves here; this is just an eternal pattern in spacetime. At 12:05 A Triangle is next to A Square; this is an eternal fact, a fact that can never change.

Try to imagine a picture like figure 137 that encompasses the entire space and time of Flatland. This vast tangle of worms and threads would make up what we call the *Flatland block universe*. You could think of making a model of the Flatland block universe by standing above Flatland and

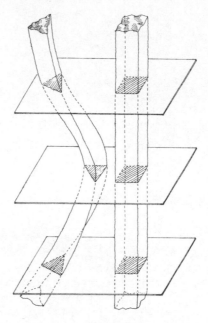

Fig. 137. A region of Flatland's spacetime.

filming the action as the polygons move around. If you then cut apart the film's frames and stacked them up in temporal order, you'd have a good model of part of the Flatland block universe.

Before going any further, I should stop to answer a question that some of you may be asking. *If we're going to think of time as a fourth dimension, does that mean that all the things we've said about the fourth dimension are really about time?* The answer is *no.* Just as there is no one fixed direction in space that we always call "width," there need be no one fixed higher dimension that is always called "time." All our talk about the fourth dimension has enabled us to think of a variety of higher dimensions: a direction in which one can jump out of space, a direction in which space is curved, a direction in which one moves to reach alternate universes. We can, if we like, insist that the past/future axis of time is *the* fourth dimension. And then we pretty well have to say that the *ana/kata* axis out of space is *the* fifth dimension, and *the* sixth dimension is the direction to other

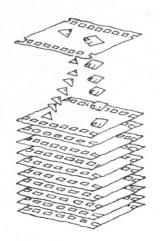

Fig. 138. Flatland's spacetime is like a stack of film frames.

curved spacetimes. But there's no point being so rigid about it. Nobody goes around saying width is *the* second dimension and height is *the* third dimension. Instead we just say that height and width are space dimensions. Rather than saying time is *the* fourth dimension, it is more natural to say that time is just one of the higher dimensions.

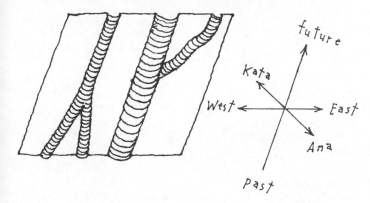

Fig. 139. *Space + Time + Hyperspace, for Lineland.*

If we think of Lineland as being a one-dimensional (east/west) world whose space bulges into hyperspace (*ana/kata*) wherever matter is present, and whose space is different at different times (past/future), then we get a 3-D space-hyperspace-time diagram as drawn in figure 139. Of course, if 1-D Lineland were beefed up to a full 3-D space, the diagram would have to be 5-D. Staying with figure 139 for just another second, note that what is shown is two segments merging to make one large segment, and an extra-large segment giving birth to a smaller segment. This picture makes up part of what we might call the *Lineland block universe.*

O.K. So now I've made the point that although time is one of the higher dimensions, there are many other possible dimensions as well. By the end of this book I'll be raving, and out-of-it, and saying space is infinite-dimensional, no doubt. But there's still a lot more to say about spacetime and the concept of the block universe.

Many philosophers argue that it is wrong to say our reality is a block universe. They do not want to represent our past-present-future universe as a static 4-D spacetime pattern.

They feel that this eternal, unchanging image leaves out something important: *the passage of time.*

Of course, the whole reason for introducing the block universe was to get rid of the passage of time. But how can I say that so universally experienced a phenomenon is nonexistent?

Wednesday, November 17, 1982

Another day has passed, and here I am trying to claim that the passage of time is an illusion. What could be more ridiculous? I remember, about five years ago, visiting my father in the hospital. He was having heart trouble, and felt depressed. I tried to cheer him up by explaining the block universe to him, and by pointing out that one's life is a permanent unchanging pattern in spacetime. "Rudy," he said wearily, "all I know about time is that you get old and then you die."

It certainly *feels* like time is passing; I'd be foolish to argue otherwise. But I want to show you that this feeling is a sort of illusion. Change is unreal. Nothing is happening. The feeling that time is passing is just that: a *feeling* that goes with being a certain sort of spacetime pattern.

Let me illustrate my thoughts with another excerpt from that imaginary classic, *The Further Adventures of A Square.*

That afternoon my Father came to advise me of my impending Arrest. Una's husband had sworn out a Warrant of Complaint. Exhausted from my morning's pleasure, I scoffed at the old Triangle's warnings and sent him on his way. What need had I to fear the vengeance of flat Polygons? Who of them could harm *me* — friend and follower of A Cube? Filled by the delicious lassitude of passion spent, I fell into a slumber.

In Dream I saw the Sphere again, floating with me in some Higher Space. His surface glowed with a solemn Luster, and I was seized with shame at my Vice. Seeking to dissemble, I called out a confident greeting.

I: Hail, lordly Sphere. Long have I sought you, long have you lingered beyond my ken.

SPHERE: The Cube would teach you on his own. Space he has taught you, but now that your Death draws near, I return to teach you Time.

I: Why speak of Death? I have not sinned!

SPHERE: Ah Square, so small is your knowing, so great is your baseness. Wouldst lie to me, who sees All? I see your past, I see your future, and your future is fraught with Peril.

I: What must I do to escape?

SPHERE: Ask the villain Cube, when next you see him. Mayhap he knows some ruse to extend your Span. But it is all foolishness, your race is Mortal. The teaching I bring to this Vision is beyond your squalid struggle for more Time. My Teaching is that Time is Unreal, and Eternity is Now.

I: What kind of Death do you foresee?

SPHERE: Silence, fool. Behold!

And there before us I seemed to see a strange and intricate Form of three dimensions. It was like a Cube, but transparent and patterned all within. Tubes and Worms and Threads ran from this strange Cube's bottom to its top; some of the Tubes were round in cross section, others were Triangular or Square. The uppermost surface of the Cube looked familiar to me, and suddenly I realized it was my World.

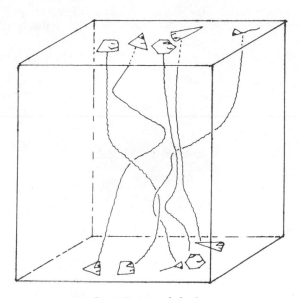

Fig. 140. A tangled tale.

There slumbered my Square form, there was my Father's familiar Triangle, and in the distance cowered a chastened Una. A Hexagon and an Isosceles were near my sleeping form, evidently bent on Violence. Only my loyal Father's intercession was keeping them at bay.

All this I saw atop the Cube. Moving my attention downward I could trace out the whole tangled history of Love and Hate.

The Isosceles's Point attracted my notice above all else, and I begged the Sphere for aid.

I: You will save me, will you not, most noble Sphere?

SPHERE: Your salvation is not mine to grant. What is this object that we gaze upon?

I: An ingenious model of some part of Flatland. There on the top is my sleeping Form, my Father, and . . .

SPHERE: What of the Cube's interior, Square?

I: You have stacked up many models of my world, O Teacher. Each plane cross section of the Cube displays a different instant of my recent (and regretted) Career. It is indeed a clever Construct, an inspired use of Higher Space.

SPHERE: Suppose I were to tell you this is no Construct? What you see is a higher level of Reality. What you see is your Space and Time. This is your World.

I: My Liege is pleased to be merry. This dead, unmoving Construct is to replace the passionate bustle of Flatland life? One could as well say that a painting breathes, or that a statue weeps!

SPHERE: The Teaching is strange, but it is no Jest. The block you see is a region of Flatland's Spacetime.

I: Spacetime, my Lord?

SPHERE: Space plus Time, thou Dullard. Hear the words of a great Spaceland thinker: *Henceforth space by itself, and time by*

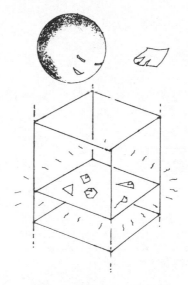

Fig. 141. The moving "Now."

itself, are doomed to fade away into mere shadows, and only a
kind of union of the two will preserve an independent reality.
Space is a shadow, Time's passing is an illusion; only Spacetime
is real.

I: Again I must protest, O Round One. Life consists of change
and Motion. Where in this Construct of Flatland's Spacetime is
there Motion?

SPHERE: You may *imagine* the Motion as follows. Suppose
that a plane were to move up through the Cube of Spacetime.
Think of the plane as a moving "Now." Fix your attention upon
it, and you will see your Form dancing its sorry Jig.

I: You are saying, then, that my conscious Mind lights up a
cross section of Spacetime, and that the passage of Time is the
upward motion of my Mind?

SPHERE: I say no such thing. There is no motion in spacetime.
Your Mind, such as it is, extends the length of your Span. More
truly spoken, the Mind is everywhere, and you have no Mind at
all.

I: I do not understand you, my Lord.

SPHERE: Nor do I understand myself.

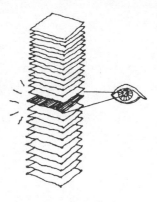

Fig. 142. Time as the motion of the mind's eye.

Thursday, November 18, 1982

The purpose of the last dialogue was to present what the
physicist David Park has called "the fallacy of the animated
Minkowski diagram." A spacetime diagram such as figure
140 is called a Minkowski diagram in honor of Russian
mathematician Hermann Minkowski (1864–1909), who
was the first to think of drawing such pictures. A Square
says that such a diagram is lacking something: the passage
of time. We do not experience childhood, adolescence, and
maturity all at once. We live through one stage, then the
next, then the next, and so on. A Square feels that figure 140
is just a *model*, but that the reality would be best represented by moving an illuminated plane up through the
spacetime solid. First one cross section would be lit up, then
the next, then the next, and so on. In this manner the static
Minkowski diagram could be *animated*, or brought to life. If
we think of the spacetime diagram as being like a movie, it
is as if A Square is saying that the movie needs to be taken
out of the can and projected. If we think of the diagram as
being like a novel, it is as if A Square is saying that the novel
needs a reader who goes through it page by page.

But there are a lot of problems with the notion of an
animated Minkowski diagram. One difficulty is that if we
think of a static spacetime, and then imagine an external

Mind, which sort of moves a searchlight up along it, we have introduced a second level of time: the time that lapses as the Mind moves its attention through spacetime. Now, if spacetime is to be *everything*, then it seems awkward and wrong to have a second kind of time lapsing external to it. With something like a novel, this poses no problem: the book incorporates its own time pattern, and the time it takes me to read the book is something else entirely. But we do not stand outside the universe like a reader outside a book; we are *in* our spacetime.

That's just my opinion, of course. I have, in fact, met people who do hold to the belief that spacetime *is* something like a novel being "read" by their soul, the "soul" being some kind of eye or observer that stands outside spacetime, slowly moving its gaze up along the time axis. I find this unsatisfying.

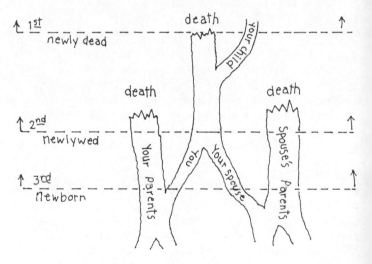

Fig. 143. Three animations of the same life.

If your life is like a novel that your soul is "reading," then the past is not so real after all. Even worse, death is just as real as ever: death is when the soul runs out of spacetime pattern to "read."

Another confusing point the animated Minkowski diagram raises is this: what if the diagram is animated more than once? What if, in other words, there is a second "reader of the novel," or a third, or a fourth, or infinitely many? What if a whole series of souls moves through your spacetime, living your life over and over? What if a whole infinite continuum of souls is moving up through your life, one at each instant at all times? But then your whole life is "lit up," so why say anything's moving at all?

Friday, November 19, 1982

What I want to say is that each of us is a certain spacetime pattern in the block universe. Today, or the day of my birth, or the day of my death — all are equally real, all are different pieces of the block universe. I will never stop living this instant. This instant will never cease to exist; this instant has always existed.

About ten years ago I had a chance to meet the great logician Kurt Gödel. I used to phone him up and ask him questions about philosophy. In my book *Infinity and the Mind* I have a section that describes a conversation I had with Gödel about the passage of time:

> We talked a little set theory, and then I asked him my last question: "What causes the illusion of the passage of time?"
>
> Gödel spoke not directly to this question, but to the question of what my question meant — that is, why anyone would even believe that there is a perceived passage of time at all.
>
> He went on to relate the getting rid of belief in the passage of time to the struggle to experience the One Mind of mysticism. Finally he said this: "The illusion of the passage of time arises from confusing the *given* with the *real*. Passage of time arises because we think of occupying different realities. In fact, we occupy only different givens. There is only one reality."

By a "given," Gödel means a person's sensations at some particular time. At any moment the world "gives" us a collection of sights, sounds, smells, and so on. By a more or less unconscious process, we organize these sensations into a stable framework. This background framework, which everyone agrees on, is reality: a continuum of three space dimensions and one time dimension. When I am in my office, I do not doubt the existence of my home. By the same token, when it is 10:00, I do not doubt the existence of 7:00. I do not think of a person as an object adrift in space; a person is a certain kind of pattern in spacetime.

A human body changes most of its atoms every few years.

The very starting point of special relativity theory consists in the discovery of a new and very astonishing property of time, namely the relativity of simultaneity, which to a large extent implies that of succession. The assertion that the events *A* and *B* are simultaneous loses its objective meaning, in so far as another observer, with the same claim to correctness, can assert that *A* and *B* are not simultaneous.

Following up the consequences of this strange state of affairs one is led to conclusions about the nature of time which are very far-reaching indeed. In short, it seems that one obtains an unequivocal proof for the view of those philosophers who, like Parmenides, Kant, and the modern idealists, deny the objectivity of change and consider change as an illusion or an appearance due to our special mode of perception. The argument runs as follows: Change becomes possible only through the lapse of time. The existence of an objective lapse of time, however, means that reality consists of an infinity of layers of 'now' which come into existence successively. But, if simultaneity is something relative in the sense just explained, reality cannot be split up into such layers in an objectively determined way. Each observer has his own set of 'nows,' and none of these various systems of layers can claim the prerogative of representing the objective lapse of time.

KURT GÖDEL,
"A Remark on the Relationship Between Relativity Theory and Idealistic Philosophy," 1959

Daily one eats and inhales billions of new atoms, daily one excretes, sheds, and breathes out billions of old ones. Physically, my present body has almost nothing in common with the body I had twenty years ago. Since I feel that I am still the same person, it must be that "I" am something other than the collection of atoms making up my body. "I" am not so much my atoms as I am the *pattern* in which my atoms are arranged. Some of the atom patterns in my brain code up certain memories; it is the continuity of these memories that gives me my sense of personal identity.

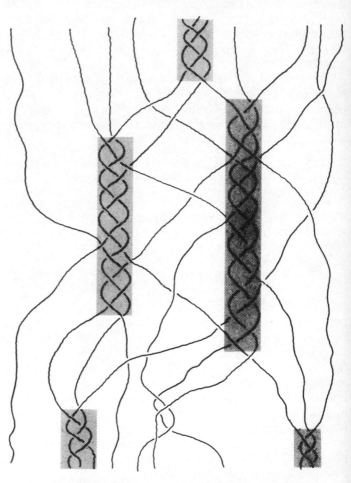

Fig. 144. People are spacetime braids of atom threads.

In figure 144 we have drawn a picture to represent the fact that people are persistent spacetime patterns. To simplify things we represent a person as a braid of three atom threads (where an "atom thread" is the spacetime trail of an individual atom). Note that in neither of the two braid-people in the picture's center do we have any single atom thread going the whole length. Note also that an atom can leave one person's braid and become part of someone else's.

What I find most striking about figure 144 is that the gray boxes enclosing individual lives are purely *imaginary*. The simple processes of eating and breathing weave all of us together into a vast four-dimensional tapestry. No matter how isolated you may sometimes feel, no matter how lonely, you are never really cut off from the whole.

I find this insight a great comfort. Instead of thinking of myself as a decaying bag of meat, I can think of myself as a part of eternal spacetime. This is a way to cheat death. Instead of identifying myself with my specific body pattern, I identify myself with the block universe as a whole. I am, as it were, an eye that the cosmos uses to look at itself. The Mind is not mine alone; the Mind is everywhere. If I do not exist, then how can I die?

Monday, November 22, 1982

"How can I die?" Well, I could have fallen off that broken viaduct I was drunkenly standing at the edge of, Saturday night, just to show off. Ow. Why, at my age, do I still do things like that? *To prove that I have free will.*

Every so often, one likes to do something foolish or unexpected. Much of life is quite predictable, but it is the crazy zigzags that give human life its peculiar savor. It's not really necessary to do something stupid and dangerous . . . Taking your wife out for dinner on a Wednesday night can be unusual enough.

If we are indeed spacetime patterns in a block universe, then the future already exists. Does this conflict with the notion that we have free will?

Not really. When I say the future is already there, I am *not* saying that the future can be predicted. When you are halfway through a detective novel, the ending is already *there*, printed on the last pages. But this does not mean that you can always guess what the ending will be. I feel that my entire life exists as a timeless whole. But this does not mean that I can predict with certainty what I'll write tomorrow, or where I'll live next year.

Sometimes when I try to explain the spacetime viewpoint to people they say, *"If you're just a pattern in spacetime and the future's already fixed, then why don't you kill yourself and get it over with? I mean, you're going to die anyway, right? Why not shoot yourself?"* The answer is simple: *"Because I don't want to."* I may enjoy dramatizing the choice between life and death by standing at the edge of a hundred-meter drop — but I'm damn careful not to fall off.

Fig. 145. I don't want to die.

It is a plant's nature to grow toward the sun, to bloom, and to bear fruit. It is a person's nature to live and love and work. In all likelihood, there is no big "Answer," and life has no significance outside itself. But that's enough. As Don Juan puts it in Carlos Castaneda's *A Separate Reality*:

> I choose to live, and to laugh, not because it matters, but because that choice is the bent of my nature . . . A man of knowledge chooses a path with heart and follows it . . . Nothing being more important than anything else, a man of knowledge chooses any act, and acts it out as if it matters to him.

The idea here is that your life is a whole, and the overall pattern is what counts. The unexpected kinks in the pattern correspond to places where you feel yourself to be making a free will decision.

Some people object very much to this view. So strong is their belief in the importance of their own free will, that they feel the future does not exist at all. They may grant that the past exists, but they feel that the block universe is something that is growing upward as time goes on. In figure 146

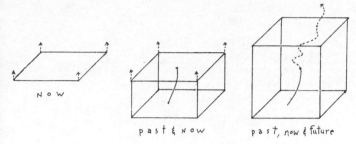

Fig. 146. Differing world views.

we illustrate this viewpoint, along with the block universe viewpoint, and the viewpoint that gives reality only to the present instant.

The great advantage the block universe has over the other viewpoints is that in the block universe there is no objectively existing "Now." Nothing is moving in the block universe, and there is no need to try to find some absolute and objective meaning for the horizontal space sheet that the other two models depend on.

As it turns out, it is actually *impossible* to find any objective and universally acceptable definition of "all of space, taken at this instant." This follows, as we shall see, from Einstein's special theory of relativity. The idea of the block universe is, thus, more than an attractive metaphysical theory. It is a well-established scientific fact.

Tuesday, November 23, 1982
Today I want to draw a lot of Minkowski diagrams: pictures of spacetime. To make things easier, we'll think of space as a one-dimensional line, and we'll think of objects as points moving back and forth on this line. The spacetime trail of a dot is called the dot's *world line.* In figure 147 we see five different sorts of world lines. *A* represents a motionless point, and *B* represents a point that moves steadily to the right. *C* is a dot that starts out motionless but then begins

---

PUZZLE 9.1
If we say that the fourth dimension is time, then it is possible to construct a hypersphere in space and time. How?

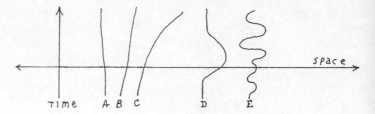

Fig. 147. Various kinds of motion.

moving faster and faster to the right. *D* makes an excursion to the right and then comes back to his starting point. *E* is in a steady state of right-left oscillation.

It is really a bit misleading to say that *A* is motionless, and *B* is moving to the right. If *A* and *B* are actual people, say astronauts floating in empty space, then all they can really be sure of is that they are moving apart from each other.

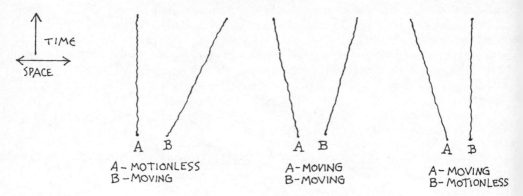

Fig. 148. Three descriptions of the same state of affairs.

Since it is impossible to make marks on the fabric of space, there really is no such thing as absolute motion. The only kind of motion one can hope to observe is the motion of one object relative to some other object. This is the content of Einstein's principle of relativity: "The laws by which the states of physical systems undergo change are not affected, whether these changes of state be referred to the one or the other of two systems of coordinates in uniform translatory motion." In formulating his theory of space and time, Einstein makes one other assumption, the principle of the constancy of the speed of light: Whenever someone measures the speed of light, he or she will always come up with the same value c ($= 29,979,245,620$ centimeters per second $\approx$ one billion miles per hour). These two assumptions have strong empirical support. Taken together, they lead to a number of surprising consequences.

Usually, in drawing Minkowski diagrams, one adopts a system of units so that the path of a light ray is represented by a 45° line. Light moves at about one billion miles per hour, so the idea is to mark off the space axis in units of billions of miles and mark off the time axis in units of one hour. In figure 149 we have marked the axes in this way, and we have drawn in the world line of a pulse of light. At time 1, A sends a light flash off to the right. One billion miles away stands M, patiently holding a mirror. After an hour's travel time, the light hits M's mirror and bounces back to the left. A gets out of the light's way, and it continues traveling to the left indefinitely.

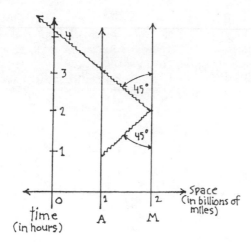

Fig. 149. World line of a photon.

As far as we know, nothing travels faster than light. It is interesting to realize that you are never actually seeing the world *right now*. What you see is always slightly in the past, as it takes the light time to get to you; what you hear is even

A person has all sorts of lags built into him, Kesey is saying. One, the most basic, is the sensory lag, the lag between the time your senses receive something and you are able to react. One-thirtieth of a second is the time it takes, if you're the most alert person alive, and most people are a lot slower than that ... We are all of us doomed to spend our lives watching a *movie* of our lives — we are always acting on what has just finished happening. It happened at least 1/30 of a second ago. We think we're in the present, but we aren't. The present we know is only a movie of the past, and we will really never be able to control the present through ordinary means. That lag has to be overcome some other way, through some kind of total breakthrough.

TOM WOLFE,
The Electric Kool-Aid Acid Test, 1968

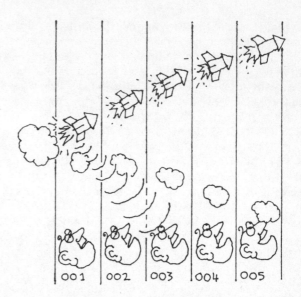

Fig. 150. Sight is faster than smell.

further in the past; and smells travel even slower than sounds. You see the sky rocket's flash, you hear its explosion, you smell the smoke. Your sensations, at any given instant, are all signals from various events in the past. Even what you feel and taste is not happening quite exactly now,

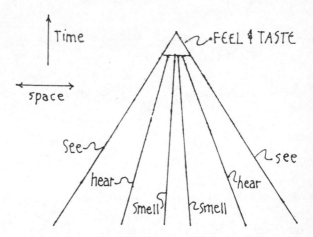

Fig. 151. One's sensations all come from the past.

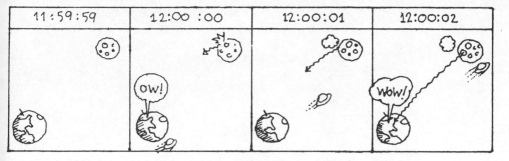

Fig. 152. *Bug Bites, UFO Zooms, Moon Booms.*

as it takes some time for nerve impulses to travel from your skin to your brain.

To talk about "all of space, taken right now," is really to speak very abstractly. You certainly don't *see* all of space right now. The tree you see is really the tree of a ten-millionth of a second ago, the moon you see is the moon of two seconds ago, the light from the setting sun started out nine minutes ago, and the twinkling stars are scattered back hundreds and thousands of years in the past. In point of fact, you can't see anything that's happening right *now*. You have to wait for the light to get to you.

It is, of course, possible for us to piece together some kind of mental image of space right now. Suppose, for instance, that it is midnight, and you are sitting outside, looking at the moon. A mosquito bites you, and then two seconds later you see the flash of some huge explosion on the moon. Since you know that it takes light two seconds to travel from moon to earth, you can conclude that the "Now" that had you being bitten by the mosquito also included the explosion on the moon.

So far this does not seem to go against any of the viewpoints that hold that there really is a spacelike "Now" moving forward through time. But there is a problem with the way in which we construct our notion of "all space taken right now." The problem is that if someone is moving relative to us, then he will construct his spacelike "Now" in a different way.

Suppose, to be specific, that a flying saucer had drifted past the Earth-moon system on that fateful night when the mosquito bit you and you saw a flash on the moon. If the aliens happen to be moving from Earth toward the moon, then they will actually reach the conclusion that the explosion happened *first*. If they are moving from the moon toward the Earth, they will, in their frame of reference, say that the explosion happened *second*. The reasoning behind these two claims is not obvious, and I won't try to give it here.

The point is simply this: differently moving observers will differ on whether or not certain events happen at the same time. Nobody can really be *at* both the mosquito bite and the explosion. Whether these two events happen at the same time is really just a matter of opinion; nature stipulates no absolute answer to the question.

The phenomenon just alluded to is called the relativity of simultaneity. The important consequence of the relativity of simultaneity is that differently moving observers will find different ways of slicing spacetime up into a stack of "Nows." The three observers drawn in figure 153 could be peacefully living in three different galaxies that are moving relative to one another. What reason could we possibly have to say that one of the three ways of slicing up spacetime is correct, and the other two are wrong?

So relativity theory implies that there is no one preferred way in which to slice spacetime into a stack of "Nows." Recall that relativity is based on the assumption that it is impossible to permanently mark a given space location. Another way of putting this is that "right here" has no real

Now and the future.

PUZZLE 9.2

What kind of ideas about the past and future are embodied in this picture, where one thinks of the spacetime solid as being something like a block of ice that melts from the bottom up?

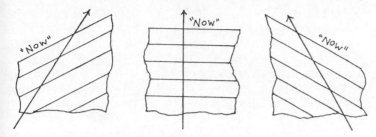

Fig. 153. Three ways to slice spacetime.

meaning over a period of time. We have seen that this assumption, oddly enough, implies that "right now" has no real meaning over a region of space. Since there is no one preferred way of defining "Now," it must be that "Now" has no objective meaning. This means that there really is no moving present, and that the block universe viewpoint is the correct one. Spacetime is a single unified whole, and the passage of time is just an illusion.

Wednesday, November 24, 1982

The special theory of relativity is extremely well tested. It makes a number of definite predictions about the behavior of matter in motion. These predictions have been borne out thousands and thousands of times in experiments involving particle accelerators.

Yesterday I was discussing one of the very attractive implications of relativity: the implication that we are eternal spacetime patterns in an unchanging block universe. But relativity also has some less pleasant consequences. The worst thing about relativity is that it implies that we can never travel faster than the speed of light.

Some people get almost paranoid about this galling prediction. Without ever bothering to learn anything about relativity, they jump to the conclusion that Einstein was simply a dogmatic spoilsport not much different from the people who used to say that man will never fly. "Why *shouldn't* we be able to go faster than a billion miles per hour? If you get a big enough rocket and run it long enough, you can go as fast as you like, right? What's the matter with Einstein anyway? Wasn't he a German or a Jew or something? Un-American for sure, and probably scared of progress!"

There are, according to relativity theory, two problems with faster-than-light travel. The first problem is that the closer any object gets to the speed of light, the greater the object's mass becomes. And the greater an object's mass is, the harder it is to accelerate the object to a yet higher speed. It would require, in point of fact, an *infinite* amount of rocket fuel to accelerate a spaceship up to the speed of light.

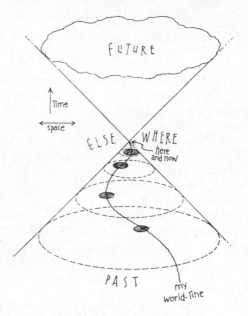

Fig. 154. Past, Future, and Elsewhere.

If I can never travel faster than the speed of light, then only a limited region of spacetime is accessible to me. Those events that I can still reach without traveling faster than light are called my "future." This is a use of the word that is peculiar to relativity theory. Normally I think of my future as being those events I will actually experience, but here I am using "future" to mean all the events that I could conceivably attend or influence, without using faster-than-light travel. In a similar vein, I can take my "past" to be all the spacetime locations from which it would have been possible to travel, in order to reach me here and now. The rest of spacetime is called "elsewhere," a lovely word. It always

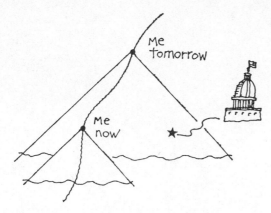

Fig. 155. "Elsewhere" today, "Past" tomorrow.

relaxes me to think about events as being "elsewhere." Whatever our leaders in Washington may be doing right this instant doesn't really matter: it's "elsewhere." Unfortunately, these important people's actions don't *stay* elsewhere — my "past" is larger at points further along my world line.

The second problem with faster-than-light travel has to do with the relativity of simultaneity. Given any event at all in your "elsewhere," there will be some observer to say it is happening at the same time as your "here and now." The whole of "elsewhere" is a sort of smeared out "Now." Usually we think of "Now" as a line between "past" and "future," but in relativity theory, this "Now" smears out into the hourglass-shaped "elsewhere." What this means is that one path into "elsewhere" is as good as any other. The trouble is that some kinds of paths into "elsewhere" seem too weird to be possible. In figure 156 I have drawn three paths into "elsewhere." The top one is unproblematic; it simply corresponds to someone who travels two billion miles per hour, or twice the speed of light. The next path is horizontal and, relative to the initial frame of reference, seems to represent a person who travels billions of miles in no time at all. This corresponds to an *infinite* velocity. The bottom path is the worst. Here the traveler seems to go backward in time. If he were to return from such a trip along a horizontal path, he would be able to meet up with his own past self! In other

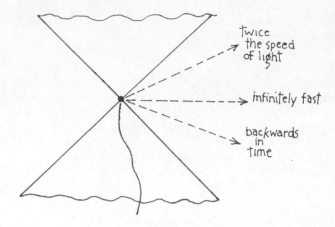

Fig. 156. Trips into Elsewhere.

PUZZLE 9.3

"The relativity of simultaneity" says that different moving observers will have different opinions about which events are simultaneous. In this problem we see how the relativity of simultaneity follows from the two basic assumptions: (1) that moving observers are free to think of themselves as being at rest, and (2) that light always travels at the same speed.

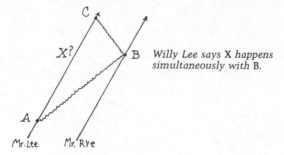

Willy Lee says X happens simultaneously with B.

The situation is as follows. A rigid platform is moving to the right at about half the speed of light. On the left end stands Mr. Willy Lee, and on the right end stands Mr. Rye. Mr. Lee sends a flash of light down the platform toward Mr. Rye. Mr. Rye holds a mirror that bounces the light flash back toward Mr. Lee. Mr. Lee receives the return signal. Call these events A, B, and C, respectively. Mr. Lee notes the times of events A and C on his world line. After a little thought he decides which event X on his world line is simultaneous with B. Where does he put X, and why? (Hint: We would place X horizontally across from B, but Mr. Lee will not. Simultaneity is relative!)

words, faster-than-light travel can lead to time travel — and many scientists feel that time travel must be impossible. The trouble with time travel is that it leads to some really vicious paradoxes, paradoxes we'll go into in the next chapter. For now, suffice it to say that so far as we know, no chains of cause and effect can proceed faster than light.

Well, that's all I have time for. I have to go home early today, to help get ready for Thanksgiving. Seems like every time you turn around it's Christmas or Thanksgiving.

Tuesday, November 30, 1982
Another holiday over with. My mother came, and my brother Embry with his wife and two kids. It was nice to see the five little cousins enjoying it: the treats, the wheel of the seasons, the eternal return.

Borges has an interesting essay on the esoteric doctrine of "the eternal return." This is the idea that *everything repeats itself.*

Many things in our ordinary experience do repeat themselves. Exhale/inhale, day/night, summer/winter, parent/child: these are cycles that repeat themselves endlessly. Obviously, my 1982 Thanksgiving is not *identical* with all the other Thanksgivings. Yet, on some level, they *are* all the same, at least as regards the broadest features, the turkey and gravy, loud children, drunkenness, quarreling, laughter, and prayer. At some point during a long holiday's welcome ordeal, most of us will experience moments, or even quarter-hours, of transcendence, of timelessness, of that full acceptance of time's fabric, which, paradoxically, sets one free from time entirely. The very repetitiousness of human rituals provides some kind of glimpse into eternity; the endless line is bent into a circle.

"The number of all the atoms which make up the world is, although excessive, finite, and as such only capable of a finite (although also excessive) number of permutations. Given an infinite length of time, the number of possible permutations must be exhausted, and the universe must repeat itself. Once again you will be born of the womb, once again your skeleton will grow, once again this page will reach your same hands, once again you will live all the hours until the hour of your incredible death." Such is the customary order of the argument, from its insipid prelude to the enormous threatening dénouement.

JORGE LUIS BORGES,
"The Doctrine of Cycles," 1934

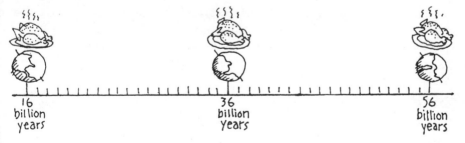

Fig. 157. The Eternal Return.

Despite all this, the eternal return is still a very unlikely notion. Can anyone really believe that at some future time the universe will be in exactly the same state as it is now? But this is what the doctrine of the eternal return asserts: there is some fixed length of time, some cosmic cycle, after which the entire history of the universe repeats itself. We might suppose a full cycle to last, say, twenty billion years.

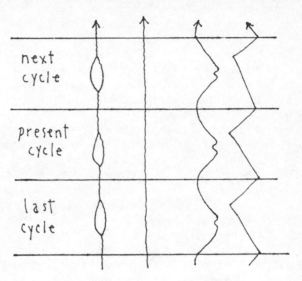

Fig. 158. Endless, repeating time.

There are two different ways of looking at a cosmos that has the property of repeating itself. We can imagine that time extends to infinity in both directions, and that space-time is made up of infinitely many identical horizontal "stripes." Alternatively, we can imagine that time is bent into a finite, yet endless, circle. If a person is strongly committed to the "moving Now" viewpoint, then the first alternative will seem more attractive. If, that is, you believe that only the present has real existence, then you will feel that even if the universe were to repeat itself, the next cycle would be somehow different. But for someone who holds to the "block universe" view of spacetime, the second way of illustrating the eternal return will seem more natural. If spacetime is indeed a solidly existing thing, then we have no real problem with bending it into a "cylinder."

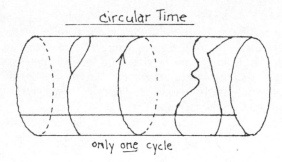

Fig. 159. Circular time.

The notion of "circular time" leads to some interesting paradoxes, which we'll discuss in puzzle 10.5. But right now let's go on and look at some of the other overall shapes that spacetime might have.

Keep in mind that in all of these pictures, we think of the horizontal axis as being space, and we think of the vertical axis as being time. The horizontal axis is, if you will, our concept of "now," and the vertical axis is our concept of "here." The event that we call "here and now" is, of course, the point where the two axes intersect. One often "reads" these spacetime diagrams by imagining the horizontal axis to move upward as time goes on. But, as I have repeatedly pointed out, it is not always necessary or desirable to animate our Minkowski diagrams in such fashion.

Figure 160 shows three kinds of spacetime. In each of them, space is infinite, stretching out endlessly in both directions; and in each of them, the future also is endless. The models differ in the way they depict the past. The picture in

See, I answer him that asketh, "What did God before He *made heaven and earth?*" I answer not as one is said to have done merrily (eluding the pressure of the question), "He was preparing hell (saith he) for pryers into mysteries." It is one thing to answer enquiries, another to make sport of enquirers.

ST. AUGUSTINE, *Confessions,* A.D. 400

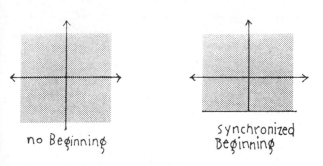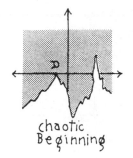

Fig. 160. Three ways to start the universe.

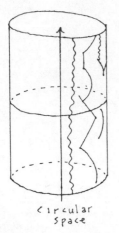

Circular
Space

Fig. 161. Circular space.

the middle shows a universe in which all of space springs into existence at once, and the left-hand picture shows a universe that has no beginning in time. The right-hand picture shows a much odder situation, a universe in which different parts of space come into existence at different times. In such a universe there might be, for instance, a big hole in space out between Neptune and Uranus, a hole that would eventually shrink and disappear. (In the picture I have labeled the event of such a hole's disappearance with the letter D.)

Another type of spacetime model arises if we assume that space is "hyperspherical," as was proposed in chapter 8. In terms of Lineland, this would mean that space is a circle, leading to the kind of picture drawn in figure 161. Figure 162 shows an alternate way of viewing a universe in which traveling far enough in any direction always leads to a region apparently identical with your starting point. Instead of showing space as curving back on itself, figure 162 shows space as extending to infinity in both directions, and with

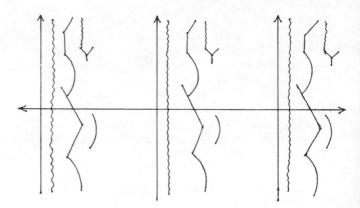

Fig. 162. Endless, repeating space.

PUZZLE 9.4

In figure 161, we drew a picture of a circular space that remains the same size as time goes on. A widely held present-day view of the universe is that our space is an expanding hypersphere, which started out as point-sized about twelve billion years ago. Can you draw a picture of spacetime that represents our space as an expanding circle?

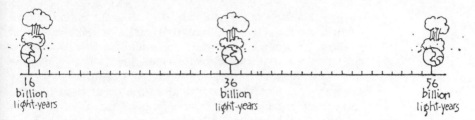

16
billion
light-years

36
billion
light-years

56
billion
light-years

Fig. 163. A pre-established harmony!

spacetime divided into infinitely many identical vertical strips. A mysterious harmony would have to be invoked to explain the repetition. Figure 161 is, of course, much more natural, just as a "circular time" model was a more natural way of representing the eternal return.

One reason that I've been dating the sections of this chapter as I write them is simply to make a little fun of my argument that time is unreal. A more serious reason is that I feel like an Arctic explorer, keeping a log of the trip as he pushes further and further into the unexplored waste. *At night the barking of the dogs, the mad splendour of the Aurora Borealis.* The goal of my quest is a certain vision, a vision of the cosmos as a pattern in infinite-dimensional

Fig. 164. Onward!

space. I want a comprehensive and absolute model for the world around me, a system that includes the realities of ordinary human life as well as the cold truths of science.

Until today I've been unsure as to whether I can finish the journey. But this morning I caught sight of the goal. Reality is one, it is a thing that provides an answer to any of the infinitely many questions one might ask; it is, in other words, a coloring-in of infinite-dimensional space. Stick with me, we'll get there yet!

10

Time Travel and Telepathy

WHY DOES TRAVEL have to be so hard? The perfect vehicle is easy to imagine: a sort of automobile with some special buttons on the dash. Get in, punch the code numbers of where and when you want to be, turn the ignition key, and — *presto* — there you are in 1920s Paris, on the Great Plains before the pioneers, on the moon, or even in another galaxy.

People have long dreamed of such a freedom from the fetters of space and time. In one Grimm brothers fairy tale, the young hero gets hold of a "wishing saddle." Get on the saddle, say where you want to be, and instantly you are there. Science fiction writers variously call this tele-portation, instantaneous matter transmission, hypertravel, or FTL travel, where *FTL* stands for *Faster-Than-Light*. Closely related is the idea of time travel, the ability to jump back to the past or forward to the future.

Will time travel and FTL travel ever become a reality? Will the final conquest of time and space ever be ours? Speaking practically, the question is what sorts of physical phenomena might conceivably make time travel and FTL travel possible. Not much is really known here, but there is some chance that by manipulating very massive systems — such as black holes — we could perhaps distort space and time in such a way as to permit the kind of spacetime leaps that time travel and FTL travel call for. Another path toward

Fig. 165. Quantum mechanics!

such travel may lead through quantum mechanics, with its hints that at the deepest level of reality, time and space do not really exist at all. If one could somehow manage to repeatedly tune in and out of the spacetime framework, one could end up anywhere and anywhen at all. But no one has any idea of how actually to do this.

After such tempting speculations, it is a little surprising to learn that most scientists reject the ideas of time travel and FTL travel. Even though no one has ever tried to carry out tests of a time machine, most scientists are confident that such tests would fail. Is this just blind prejudice?

Not really. The problem with time travel is that it leads to physical paradoxes, to contradictions in the fabric of reality. And most scientists feel that our world is too logically put together to allow the occurrence of direct contradictions. The reasons for rejecting time travel and FTL travel are thus of an a priori nature: *Contradictions cannot occur in the world, time travel and FTL travel can lead to contradictions, therefore there is no such thing as time travel or FTL travel in our world.*

This is a subtle argument, and well worth considering in some detail. First of all, what is really wrong with having a contradiction in the world? Don't contradictions happen all the time? I want a hamburger, and I don't want a hamburger. A photon is a particle, and a photon is a wave. A zebra is white, and a zebra is black.

These are contradictions of a sort, but not really unbearable ones. The fact that I both want and don't want a hamburger points only to the fact that "I" am a conglomerate of conflicting desires. A photon is never observed to be both particle and wave at the same instant. A zebra is black and white, but in different stripes.

Fig. 166. Yes and no.

Although our world may appear to have some contradictions in it, these contradictions can ordinarily be resolved by making a finer distinction. But what about an absolute contradiction? What about some concrete, specific fact *A* for which both *A* and not-*A* are true? Here are two examples of what I call *yes-and-no paradoxes* in time travel.

1. At age thirty-six, Professor Zone suffers a temporary psychosis. During his period of madness, he murders his beloved wife Zenobia. He is found not guilty by reason of insanity, but, stricken with remorse, he decides to devote all his energies to undoing his wrong. He hopes' somehow to go back and change the past. On his fiftieth birthday, Zone finally completes his work: the construction of a working time machine. He gets in the machine, travels back some fourteen years, and goes to look in the window of the house where he and his dear wife used to live. There is his poor Zenobia, and there is that mad killer, Zone-36. Zone-50 had hoped to arrive early enough to talk some sense into Zone-36, but the crucial moment is already at hand! Zone-36 is stalking Zenobia, a heavy pipe wrench raised high overhead! Without stopping to think, Zone-50 aims his bazooka and shoots mad Zone-36 through the heart. *The Paradox*: If

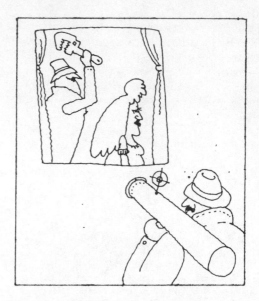

Fig. 167. Professor Zone.

Zone-36 dies, then there can be no Zone-50 to come back and kill Zone-36. If Zone-36 does not die, then there will be a Zone-50 to come back and kill Zone-36. Does Zone-36 die? *Yes and no.*

2. On Monday the Beagle Boys steal Uncle Scrooge's new time machine. They use it to pop into the future and find out who will win the big horse race on Wednesday. It'll be Ole Plug, a hundred-to-one shot! Back on Monday they wonder how to get enough money to bet. All they have is one lousy dollar, and the minimum bet is two bucks. "I've got it." Beagle Boy 22-03-46 grins. "On Thursday we'll send one of our two hundred greenbacks back to Tuesday. That way on Wednesday we'll have the two smackers to put on Ole Plug!" Tuesday, sure enough, a dollar bill appears in their hideout. Wednesday Ole Plug wins, just as they knew he would, and the Beagle Boys all go out to celebrate. Unfortunately, they overdo it a bit, and on Thursday they're down to one lousy dollar again. "O.K.," says 22-03-46, "time to send this dollar back to Tuesday!" "Forget it," answers young 23-08-69. "I'm taking this dollar to buy a can of beans." 23-08-69 snatches the bill away from 22-03-46, and leaves. *The*

Paradox: Since the dollar appeared Tuesday, they must have sent it back from Thursday. Yet, when the time comes, they *don't* send it back. Do they send a dollar back? *Yes and no.*

These paradoxes are, on one level, little more than amusing intellectual games. One tends to feel there is always some way to weasel out. What if Zone-50 actually went to the wrong house? What if 22-03-46, worried about the paradox, manages to get a dollar on Friday and *then* sends it to the Tuesday Beagle Boys? But you can't always be sure of lucking out. Time travel can lead to irreconcilable paradoxes. I'd like now to present a nice, clean version of such a paradox, in terms of a Minkowski diagram.

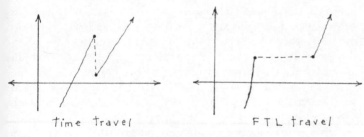

Fig. 168. Two SF dreams.

Keep in mind that in these spacetime diagrams, the horizontal direction is space (Lineland, strictly speaking), and the vertical direction is time. World lines involving time travel and FTL travel would look something like the diagrams in figure 168. In each case I have used a dotted line for the part of the world line that represents the "jump." I do this to suggest that such jumps, if at all possible, would probably be done by somehow moving out of normal spacetime to travel through some higher dimension. A second point worth making here is that if you are going to time-travel to your own past, it is wise to do it while you are in motion. Otherwise you may jump to a place occupied by your past self, and there could be a nasty explosion. So the time traveler in figure 168 is moving slowly to the right both before and after his jump. (We will assume that the time machine jumps back to the same position, *relative to the laboratory*, that it jumps from.)

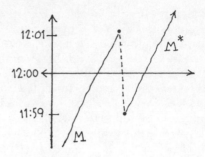

Fig. 169. A two-minute jump back.

Now for the paradox. Suppose that I build a small time machine capable of transporting itself two minutes back in time. Around 11:55 A.M. I set it slowly rolling to the right on my laboratory bench, and with a timer switch set to initiate the jump at 12:01. I sit there watching, and at 11:59 there are suddenly two time machines on my bench: *M*, the one that has not yet jumped, and *M**, the one that has jumped back from the future. For two minutes both machines are there, and then at 12:01 the timer switch goes off and *M* disappears. After 12:01 I am left only with *M**, which is really a later version of *M*.

Fig. 170. Two time machines.

So far, so good. Now we introduce the paradox. Suppose that for safety reasons my time machine *M* is equipped with a sonar device to make sure that the laboratory bench is clear before any jump takes place. If *M* senses any other object on the bench with it at 12:01, then it overrides the timer switch and refuses to jump. Now repeat the experiment. What happens?

If M^* appears at 11:59, then it will still be around at 12:01, and M will sense it with its sonar. If M senses M^*, then M will refuse to make the jump. And if no jump takes place, then M^* does not appear at 11:59.

If M^* does not appear at 11:59, then the bench will be clear at 12:01, and M will make the jump as planned. If M makes the jump, then M^* appears at 11:59.

Conclusion? M^* appears at 11:59 if and only if M^* does not appear at 11:59. Now, one of the two alternatives must actually happen: either M^* appears or it doesn't. But we have just proved that if either alternative happens, then the other alternative happens as well. So M^* appears on the laboratory bench at 11:59, and M^* does not appear on the laboratory bench at 11:59. Does M^* appear? *Yes and no.*

It is very hard to imagine a world in which such a logically contradictory state of affairs is possible. Since the existence of a time machine can lead to such a contradiction, many scientists feel that time machines are logically impossible.

Further, it is possible to show that any FTL travel machine can be adapted to become a time machine. The argument, which we briefly mentioned in the last chapter, hinges on what Einstein dubbed the relativity of simultaneity. Loosely speaking, the idea is that once you travel faster than light, you are in fact traveling into the past relative to some observers. Once this is accomplished, you can change your speed in such a way as to end up in your own past. In other words, an FTL traveler can return from his trip before he leaves — and this is time travel.

Since FTL travel leads to time travel, and time travel leads to logical contradiction, many scientists feel it is also possible to rule out FTL travel by a priori reasoning.

How strong are these arguments, really? Returning to our "disproof" of time travel, there seem to be three kinds of loopholes. (1) What if there were time machines, but no one ever used them to produce contradictions? Perhaps some kind of "Time Police" might be recruited to prevent such experiments. Or maybe the cosmos would, in the interest of self-preservation, strike dead anyone about to perform a paradoxical time travel experiment! (2) Maybe there can actually be contradictions in the fabric of reality. It is, perhaps, not *completely* impossible to think of an M^* that both ap-

I suppose a suicide who holds a pistol to his skull feels much the same wonder at what will come next as I felt then. I took the starting lever in one hand and the stopping one in the other, pressed the first, and almost immediately the second. I seemed to reel; I felt a nightmare sensation of falling; and, looking round, I saw the laboratory exactly as before. Had anything happened? For a moment I suspected that my intellect had tricked me. Then I noted the clock. A moment before, as it seemed, it had stood at a minute or so past ten; now it was nearly half-past three!

I drew a breath, set my teeth, gripped the starting lever with both hands, and went off with a thud. The laboratory got hazy and went dark. Mrs. Watchett came in, and walked, apparently without seeing me, towards the garden door. I suppose it took her a minute or so to traverse the place, but to me she seemed to shoot across the room like a rocket. I pressed the lever over to its extreme position. The night came like the turning out of a lamp, and in another moment came tomorrow. The laboratory grew faint and hazy, then fainter and ever fainter. Tomorrow night came black, then day again, night again, day again, faster and faster still. An eddying murmur filled my ears, and a strange, dumb confusedness descended on my mind.

H. G. WELLS,
The Time Machine, 1895

pears and does not appear. Maybe contradictions are rare, but not totally ruled out. After all, there is a sense in which the very existence of our world is a contradiction: for how could something come from nothing? (3) Perhaps there is some more refined sense of the word *exist* under which something could manage both to exist and not exist. If there were many parallel universes, for instance, then we could have M^* existing in some and not existing in others. The simplest solution along these lines would be to postulate that a time machine always travels to the past of some world other than the world it starts out in. Paradoxes arise only if you go into your *own* past and do something like smothering your own grandfather in his cradle. If you kill some poor baby in a parallel world, no contradiction comes up.

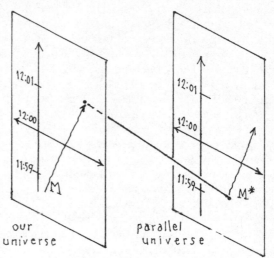

Fig. 171. "Time travel" to another universe.

If we redraw our two-minute time machine paradox in terms of parallel worlds we get figure 171. When the machine jumps back in time, it also jumps over to a different sheet of spacetime, as indicated. In a situation like this, all you would see would be the disappearance of your "time machine" at 12:01. You'd probably never see the machine again — though if it was programmed to do further jumps it might conceivably reappear in your laboratory at some future time. Most science fiction writers use this notion of parallel universes to avoid the paradoxes of time travel. Strictly speaking, of course, travel to a parallel world is not time travel at all.

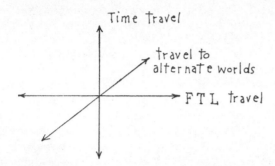

Fig. 172. The three freedoms.

It is interesting to put into one diagram the three kinds of "special travel" that appear over and over in science fiction tales: time travel, travel to alternate worlds, and FTL travel. These correspond to three mutually perpendicular types of motion. The great popular appeal of these kinds of travel is that they promise freedom from the fetters of the human condition. Time travel sets one free from the blind juggernaut of time, free from fruitless nostalgia. FTL travel frees one from the obstinate tyranny of physical distance, from the dull necessities of actual travel. Travel to alternate worlds frees one from having to occupy any fixed position in society, and frees one from having to accept the world as it is. At the deepest level there is really not that much psychological difference among the three sorts of travel; each provides a magical escape from the here-and-now-and-how. Rationally, we all know that we can change our lives if we really want to: you take a vacation, you find a new job, you sell your house and move. But making a big change is so hard. How much easier it would be to just get in a machine and push some buttons!

So far we have discussed only one basic type of time travel paradox, the yes-and-no paradox. There is another, less vicious kind of time travel paradox, the *closed causal loop*. Here are two examples.

Fig. 173. Which came first?

174 · How To Get There

The instructions were on the back of the clasp — when I touched it lightly, the words TIMEBELT-TEMPORAL TRANSPORT DEVICE winked out and the first "page" of direction appeared in their place. Every time I tapped it after that, a new page appeared. They were written in a kind of linguistic shorthand, but they were complete. The table of contents ran on for several pages itself:

OPERATION OF THE TIMEBELT
  Understanding
  Theory and Relations
  Time Tracking
  The Paradox Paradox
  Alternity
  Discoursing
  Protections
  Corrections
  Tangling and Excising
  Excising with Records
  Reluctances
  Avoidances and Responsibilities

FUNCTIONS
  Layout and Controls
  Settings
  Compound Settings
  High-Order Programming
  Safety Features

1. An inventor is in his laboratory, struggling to assemble a working time machine. Suddenly there is a flash of light, and a man from the future appears, riding a lovely time machine. "I'm an historian," says the man from the future. "I want to interview you, as you are the inventor of the time machine." "But I don't know how to build one yet," replies the inventor. "I don't know if I'll *ever* get it right." "Well, here," says the helpful historian, "just look my time machine over, and build yourself a copy of it." *Who invents the time machine?*

2. In 1969 the childless yet kindly Goodcheese couple find a baby girl on their doorstep. They name her Cynthia and raise her like a daughter. Cynthia shows an incredible aptitude for physics and earns her Ph.D. from Cal Tech at age nineteen. She falls in love with Randy Crassman, a young biologist deeply involved with cloning research. Foundations shower money upon them. Cynthia constructs the world's first time machine, and Randy manages to get one of Cynthia's cells to grow into an exact copy of baby Cynthia herself. A conservative faction takes over the government and rules that the cloned baby must be destroyed. Tearfully, Cynthia puts the baby into her time machine and sends it back to 1969. The baby, as chance would have it, lights on the doorstep of the childless yet kindly Goodcheese couple. *Where did Cynthia really come from?*

A simple "laboratory" example of a closed causal loop would be the following. One morning I come into my laboratory and putter around, cleaning off my workbench. At 11:59, to my surprise, a small two-minute time machine appears on the bench. To test if it really works, I set it to jump back two minutes at 12:01. At 12:01 it disappears.

In figure 174 we can see the loop very clearly. There's no actual contradiction here, but it's certainly a weird situation. At first one may be tempted to think of the little time machine as circling around and around the loop. This temptation should be resisted! If we stick to the spacetime viewpoint, we do not have to imagine that anything in figure 174 is actually *moving*. There is simply a circular loop here, a circle with no beginning and no end.

Such closed causal loops are not ruled out in modern physics. Far from it! According to quantum mechanics, empty space actually seethes with little matter / antimatter loops. The idea is that energy, such as is carried by a photon

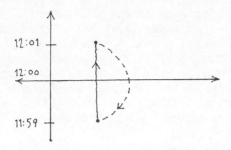

Fig. 174. A closed causal loop.

DAVID GERROLD,
The Man Who Folded Himself,
1973

of light, can be briefly converted into mass, and then reconverted back into energy. At a given point, one might have an electron and a positron emerging out of nothing, only to bump back into each other and disappear.

The reason we might think of this as a closed causal loop is that a positron is sometimes thought of as an electron that goes backward in time. A *positron*, I should explain, is a particle with exactly the same mass, spin, size, and so on, as an electron. The only difference between the two is that the electron carries a charge of *minus* one, and the positron carries a charge of *plus* one. These matched particles are said to be a matter / antimatter pair because whenever an electron and a positron get close to each other they disappear in a flash of light. This process is called *mutual annihilation*. The other side of the coin is that whenever you create an electron out of nothing, you also have to create a positron at the same time. This process is called *pair production*.

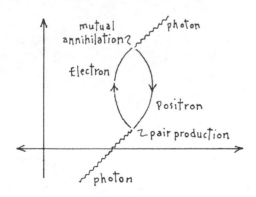

Fig. 175. A pattern in the mass-energy dance.

The kind of process illustrated in figure 175 is known to happen very commonly. If we look at it in terms of the "moving Now" viewpoint, it seems a little surprising that the electron and positron manage so neatly to appear and disappear together. But according to physicist Richard Feynman, one can take a spacetime viewpoint and regard the positron as an electron that is traveling backward in time. From this standpoint we simply have a nice little closed causal loop.

The notion of a particle "moving backward in time" is not really taken too seriously by the physicists. It is more a mathematical fiction than an actual phenomenon. No physicist, as far as I know, entertains hopes of using a beam of positrons to somehow send signals into the past. But once you start thinking in these terms it is hard to stop. What would it be like to actually live part of one's life backward in time? A few years ago I wrote a very short story in which the heroine goes around a "corner in time." Here it is, complete with Minkowski diagram:

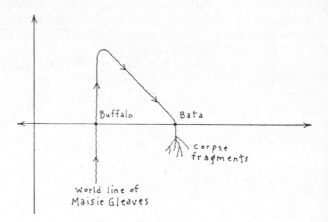

Fig. 176. Diagram for "A New Experiment with Time."

A NEW EXPERIMENT WITH TIME

The first thing the citizens of Bata notice is a greasy place in the street. A fat man slips on it. Bill Stook comes down in the yellow pickup with the smashed fender and throws on a bucket of sand.

A week later the patch begins to stink. The stuff has thickened and drawn together. There's lots of flies that come and land on your face afterwards. The kindergarten teacher twists her ankle. Black high-heels and a thin summer dress.

Stook comes back with a shovel, but he can't get the stuff loose. A few idlers — daytripping feebs — give their advice, spit-talking about glare-ice and mineral oil. Finally Stook throws on some more sand and goes home.

Under the arc-lights the patch is elliptical, four by eight feet. It cuts across the crosswalk and both lanes of traffic. The tire-marks on it extend out into straight smears in either direction. A dog has dropped a bone in the middle.

Maisie Gleaves lives in a Buffalo rooming house. She is black and white, with red lipstick and a christmas-green raincoat. Every night she lies on her bed, looking at her Bata High School yearbook. Two years now. Somehow she will go back.

Workmen are putting up a banner saying, BATA SIDEWALK SALE DAYS. Meanwhile a group of men, shopkeepers, inspect the stinking patch of pulp. One of them tries to pick up a bone. His fingers slide off it. It's an outrage. Bill Stook is called and threatened with dismissal.

He covers the patch with sawdust and puts a refreshment stand around it. SIDEWALK SALE DAYS. In the hot sun, people order hot dogs, catch a whiff of decay and put on more mustard. Stook mans the booth, nipping whiskey from a pint bottle. The flattened lump underfoot feels springy.

A white sunset slides under low clouds. They dismantle the booth and the sawdust blows away. Mashed arms and legs, tooth cracklings, scraps of green cloth. The tire tracks are gone from the flattened corpse. The state police take Stook away.

Maisie watches Buffalo TV in a silver diner. Trouble in Bata. She remembers all the lost faces. Ron. She pays for her tea. Back in her room she stares into the mirror for two hours. Her image is moving closer.

Sleeping or waking, it's all the same now. No more boundaries. Something is coming nearer, growing to connect. She lives on air and thinks only of Bata. She will return.

Bit by bit the corpse grows whole. Slowly the bones link up, imperceptibly the flesh crawls back. One night the face is finished. In the dark is begins to twitch unseen.

Stook is out on bail. He is driving a stolen truck, the pickup they used to let him use. All his rage and bitterness is focused on the corpse in the street. He speeds towards it, past the guards, through the sawhorses. A screech of brakes, a thud. Suddenly his crumpled fender is smooth. The corpse walks off backwards.

Let us imagine an Intelligence who would know at a given instant of time all forces acting in nature and the position of all things of which the world consists; let us assume, further, that this Intelligence would be capable of subjecting all these data to mathematical analysis. Then it could derive a result that would embrace in one and the same formula the motion of the largest bodies in the universe and of the lightest atoms. Nothing would be uncertain for this Intelligence. The past and the future would be present to its eyes.

PIERRE-SIMON LAPLACE,
Theory of Probability, 1812

Stook runs after the skinny corpse, a woman. She minces backwards towards the bus stop, glaring at him. He catches up as she climbs into the bus to Buffalo. He tries to grab her, but it's impossible. He cannot alter her past.

Maisie leaves her room and walks. A block ahead she sees a black and white woman in a christmas-green coat climb off the bus from Bata. She is walking backwards, this woman. Maisie hastens to meet her.

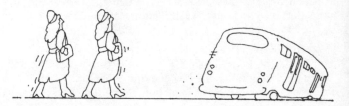

Fig. 177. Maisie Gleaves.

The two figures merge and are no more. A cabbie sees them disappear into each other. For Maisie it is different. She walks through the flash and down the street.

Everything is running backwards. Maisie is going back through time, back to Bata. The bus backs up to where she'd seen herself get out. Ticketless, she climbs in the exit door and sits down. She is nervous. The bus is going forty miles per hour in reverse.

As the bus backs out of Buffalo onto the Thruway, the man sitting next to Maisie begins staring at her. He says something backwards, a drooling gabble. She answers anyway. He turns and stares out the dark window. She spoke because he spoke; he spoke because she spoke. He picks off a wad of gum from under his seat and begins chewing it.

When the bus leaves the Thruway and backs past the old filling-station, she walks to the door. It opens, and she goes down the steps. Bata. She's glad she waited so long. She'll get a room here, and in two years she'll be back in high-school. Ron. This time it will work out right.

A short, red-faced man is blocking her way. She sets her face and walks towards him. He backs off, drawing farther and farther away. There are police around a pickup parked in the intersection. But there is no traffic.

The little man scuttles crablike into the cab of the pickup. Just to scare him, she walks right up to it, right up to the fender. There is a sudden jolt. The pickup squeals its brakes and backs away.

The story was designed as a sort of thought experiment. The idea is that a universe that has some spacetime patterns other than ordinary cause and effect is at least *conceivable*. We cannot very easily imagine a mechanism that would actually make Maisie Gleaves turn a corner in time, but a spacetime that has such a corner is still comprehensible to us. For the rest of this chapter I want to discuss the idea that our spacetime may actually have patterns other than those introduced by simple cause and effect.

In the nineteenth century, people used sometimes to think of the universe as being like a big clock: *Many, many years ago, God the big watchmaker put the whole thing together, wound it up, and walked off. Given full knowledge of the clockwork universe at any time, the whole past and future can be extrapolated.* This is really a very limiting and lifeless way of thinking of the world. The one thing that may have made the clockwork universe attractive to people is that once God finishes putting it together, you don't have to have Him hanging around anymore, meddling with your affairs.

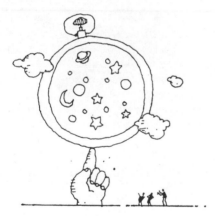

Fig. 178. The clockwork universe.

A great weakness with the clockwork universe viewpoint is that the world does not in fact seem to behave very deterministically. That is to say, a given cause does not always produce the same effects. Even if you pull the plunger back to the exact same point, two different balls are likely to

produce totally different scores on a pinball machine. For
many years, advocates of the clockwork universe ascribed
such variations to invisibly small motions of atoms. Underlying the world's seeming disorder was supposed to be a
completely lawlike mechanism of bouncing atoms. But
with the introduction of quantum mechanics in the twen-
tieth century, we have come to regard the behavior of even
the simplest atom as essentially unpredictable.

Fig. 179. Essentially unpredictable.

Cause and effect still tie together many of the events in
our world. But it is also true that many events happen for no
real reason at all. If there were a God directing the world'
history, then it would not have been enough for him to se
up the world billions of years ago and then walk off. He
would have to work over all of spacetime, weaving the ran
dom blips into a pleasing pattern.

I certainly do not want to put myself in the position o
arguing for the existence of some giant humanoid God. The
point I am really trying to get at is that since the process o
cause and effect does not in fact account for all the world'

PUZZLE 10.1

*If your were to have complete freedom in moving forward and backward in time, then
you could duplicate most of the feats of a hyperbeing who can move ana and kata at will.
How could you use time travel to enter a sealed room? How could you use it to remove
someone's dinner from his stomach without disturbing him?*

Fig. 180. Synchronicity.

structure, we might well look for other kinds of patterning. I am thinking specifically of the acausal connecting principle known as *synchronicity*. What is synchronicity?

You learn a new word, and during the next week you see it in three different places. You're thinking of an old high school friend you haven't seen in years, the phone rings, and it's your old friend on the line. You go to a convention, hoping to have a discussion with a certain Dr. X . . . and who should be sitting next to you on the bus in from the airport but Dr. X himself.

Meaningful coincidences. Life is full of them. Where do they come from? What do they mean? Can they be controlled?

C. G. Jung, the great Swiss psychologist, began using the term *synchronicity* in 1920, to mean "acausal connection" or "meaningful coincidence." His clearest statement of his ideas about synchronicity is to be found in his introduction to the *I Ching*, an ancient Chinese book of divination. To consult the *I Ching*, one flips a group of three coins a total of six times to generate a "hexagram," a pattern of solid and broken lines. There are sixty-four possible hexagrams, each of which bears a corresponding name and page of advice.

Synchronicity is not a philosophical view but an empirical concept which postulates an intellectually necessary principle. This cannot be called either materialism or metaphysics . . .

Synchronicity is no more baffling or mysterious than the discontinuities of physics. It is only the ingrained belief in the sovereign power of causality that creates intellectual difficulties and makes it appear unthinkable that causeless events exist or could ever occur. But if they do, then we must regard them as *creative acts*, as the continuous creation of a pattern that exists from all eternity, repeats itself sporadically, and is not derivable from any known antecedents . . .

Meaningful coincidences are thinkable as pure chance. But the more they multiply and the greater and more exact the correspondence is, the more their probability sinks and their unthinkability increases, until they can no longer be regarded as pure chance but, for a lack of causal explanation have to be thought of as meaningful arrangements.

C. G. JUNG,
Synchronicity, 1952

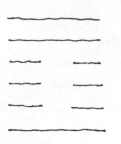

Fig. 181. The hexagram for "Increase."

The really surprising thing about the *I Ching* is how often the advice it dispenses is precisely and unambiguously relevant to the questioner's actual situation.

An example: a young married couple of my acquaintance were expecting an unplanned third child. They decided they could not afford it, and scheduled an abortion. Doubts over this course of action set in, and they consulted the *I Ching*. The reading? "Increase. It furthers one to undertake something. It furthers one to cross the great water." Heeding this advice, they braved the waters of birth once again, and had their third baby. What made this story truly striking to me was that shortly thereafter Alison Lurie's novel *The War Between the Tates* appeared . . . and this novel contains a description of a couple who go through the same sequence of events.

Statistically, of course, one expects a certain number of remarkable coincidences to occur. If you play roulette long enough, there will come an evening when your every bet wins. Usually when I think of someone, he does not in fact phone me up . . . So is the one time that the call *does* come so very surprising?

Fig. 182. Not all premonitions are correct.

There is also a certain amount of self-suggestion present in the observation of coincidences. If, for instance, my arm is in a cast, I will be very conscious of casts . . . I will see people in casts everywhere I go. Is this synchronicity, or is it just a simple alteration in the way I am seeing the world?

Fig. 183. You notice people who look like you.

Probably there were always a lot of people in casts around, but I just didn't notice them.

Yet this last example does seem to contain the germ of an explanation of synchronicity. According to quantum mechanics, reality as we know it is the product of a mutual interaction between the objective world and the subjective observers. Jung views such interactions as fundamental to synchronicity: "Synchronicity takes the coincidence of events in space and time as meaning something more than mere chance, namely, a peculiar interdependence of objective events among themselves as well as with the subjective (psychic) states of the observer or observers." Does an individual's frame of mind determine what happens to him or her?

There is a very pretty illustration of synchronicity in Steven Spielberg's famous movie *E.T.* E.T. (for Extra-Terrestrial) is a pleasant, goblinlike creature who is befriended by a ten-year-old boy named Elliot. E.T. seems to have great psychic powers, and his mind becomes in some way linked to Elliot's. At one point, E.T. is watching a romantic adventure on television, while Elliot is at school fighting with a girl whom he admires. Suddenly the two images — Elliot-and-girl, hero-and-heroine — take on the same Gestalt, and just as the hero kisses the heroine on TV, Elliot kisses the girl in his classroom. Is it that E.T.'s perception of the TV show alters Elliot's reality? Or would it be better to say that Elliot and E.T. are part of a linked system that manifests the same experience in different ways?

The first way of putting it suggests that E.T. has some sort of "mind ray," which, by a chain of cause and effect, alters what is happening in Elliot's classroom. The second way of looking at the coincidence suggests, rather, that there is some sort of noncausal harmony between the actions of Elliot and E.T.

In the case of this totally fictional movie, it is, of course, the second view that is correct. The child actor portraying Elliot and the roboticized doll representing E.T. have the same "experience" because Steven Spielberg, the movie's creator, intended them to. He *designed* the movie so as to contain this particular synchronicity.

Movies and novels very commonly include large numbers of significant coincidences. These synchronistic elements provide a sort of "horizontal" structure complementary to the "vertical" development of the plot. By the same token, one might think of our world's synchronicities as a sort of "horizontal" spacetime structuring, a structuring that is there for reasons that might be thought of as artistic.

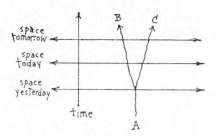

Fig. 184. A common cause.

Let me make this "horizontal versus vertical" distinction a bit clearer by means of two spacetime diagrams. In figure 184, we have a spacetime diagram representing an object A, which yesterday split into two objects B and C, which are now drawing apart from each other. We might, for instance, take A to be an amoeba which splits into two identical amoebas B and C. B and C are identical, but this is not an example of synchronicity. The similar features of B and C can be traced back to a common cause, their parent A.

Figure 185 illustrates what might be a synchronistic event. Here we have the two distant objects B and C coinci-

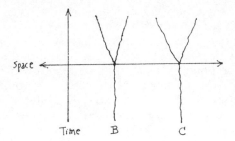

Fig. 185. A coincidence.

dentally splitting up at the exact same time. B's split is not the cause of C's, and C's is not the cause of B's. They just happen to occur at the same time.

The point is that cause and effect can be regarded as a sort of "vertical" patterning of spacetime, while synchronicity is a "horizontal" patterning of spacetime. Cause and effect set up certain branching patterns in time; synchronicity arranges these patterns in step with each other. When both patternings are at work, one gets the kind of complex pattern of events characteristic of life as it is lived.

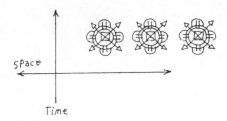

Fig. 186. Cause-effect and synchronicity.

It seems evident that a really first-class universe must include a mixture of both sorts of spacetime patterning. What I am suggesting, in short, is that our world contains synchronicity because it is a beautiful and interesting world! But who put all that synchronicity there? Who built in all the deep meaning?

Many would answer that God did it. Indeed, one of the three traditional theological proofs of the existence of God is the "argument from design," which reasons that the universe, as a supreme work of art, must have been made by some great Artist. But this conclusion is not inescapable. According to the quantum-mechanical world view, it is we ourselves who create the world, event by event, instant by instant. The universe is, in a sense, a book written by its characters, a dream dreamed by its own phantoms.

But why, then, should the universe exhibit such overall coherence, such synchronicity? Curiously enough, quantum mechanics not only allows for synchronistic events . . . It *requires* them. This result, established as recently as the 1970s, has its original inspiration in the so-called Einstein-Podolsky-Rosen paradox of 1935.

The EPR paradox goes as follows. Quantum mechanics predicts that once two particles have been near each other, they continue to instantaneously affect each other no matter how widely they may be separated. Now we know from Einsteinian relativity that no signals can travel faster than light. Returning to figure 185, we can see that there is no way a sudden change in B can *cause* a synchronized change in C (or vice versa), since there is no way for a signal to instantaneously zip from the one to the other. In Einstein's view this was a paradox. B and C act in harmony, even though they cannot exchange signals fast enough to causally *organize* their harmony.

Einstein was, in some ways, a very deterministic thinker. He viewed synchronicity as too miraculous a phenomenon to be built into our physics. He sought to escape the conclusions of the EPR paradox by postulating the existence of "hidden variables," tiny internal clocks, as it were, that would be the hidden cause of B and C's simultaneous decay.

Now, in the case where A, B, and C are *amoebas*, such a view is correct. If A has parented B and C, then the fact that B and C mature simultaneously is not really synchronistic. Two flowers blooming at the same time is not synchronicity; it is evidence, rather, of a common hidden cause: the

---

PUZZLE 10.2

Special relativity says that it is impossible to permanently label any given space location. "Right here last week," in other words, has no absolute meaning. How would the existence of a time machine go against this assumption?

plants' common ancestor with its built-in biological clock. The amoebas B and C mature in concert because they have the same DNA. Einstein proposed that even if B and C were things as simple as photons or electrons, then there would still be some internal structure, some "hidden variables," to account for their synchronistic behavior.

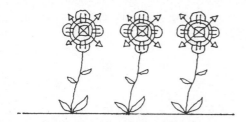

Fig. 187. A hidden common cause.

These hidden variables cannot be directly observed. But in 1964, physicist John S. Bell devised a type of experiment that could test for these hidden variables in a statistical way. In the 1970s, physicists at Berkeley, Harvard, and other universities carried out a series of such experiments verifying what the quantum physicists had said all along: there are no

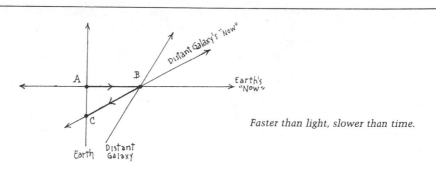

Faster than light, slower than time.

PUZZLE 10.3

Here is a picture illustrating how FTL travel can lead to time travel to one's own past. The traveler goes from A to B to C, B being an event on the world line of a distant galaxy that is moving away from Earth at half the speed of light. Explain how the paths AB and BC both can be regarded as pure FTL trips.

hidden variables, even though it is true that distant particles can behave in a synchronized way.

This result rules out the existence of the sorts of hidden causes that Einstein desired. An electron is just what it seems: a completely simple particle with no memory, no internal clock, no "hidden variables." The fact that two widely separated elementary particles B and C can act in concert *has no explanation*.

This, finally, is the essence of synchronicity: the world we live in is filled with harmonies and coincidences that have no explanation in terms of cause and effect. It is fruitless to seek after hidden forces and occult powers. The world is a given — it is just as it is, full of cause and effect, full of synchronicity.

Some people have taken these new developments to mean that such psychic feats as telepathy and psychokinesis are now firmly established scientific realities. This is not correct. To understand this point, we should make clear what is meant by telepathy and psychokinesis (PK for short).

The basic supposition behind these ideas is that if I am a psychic, then by thinking a certain way I can change the world around me. In the case of telepathy, I am to transmit my thoughts to other people; in the case of PK, I am to make objects move.

A very obvious fact — which is somehow overlooked in most discussions of psychic phenomena — is that, in a sense, we all have telepathy and PK. I have some thoughts in my head, I go talk to you, and then you have the same thoughts! I want certain words to appear on my typing paper, I will my fingers to move, and the words appear! I want my body to move into the next room, I move my legs, and my body is in the next room! There is a sense in which these are all rather amazing phenomena. Who needs telepathy when we have telephones? Why wish for PK when you have hands? What makes teleportation so much better than airplanes?

PUZZLE 10.4

Not only does FTL travel lead to time travel; the converse is true as well. Time travel leads to FTL travel. Given a rocket and a time machine, how could you send a probe all the way around the galaxy, yet have it get back the same day!

But of course people would like to have more. A somewhat outdated paradigm for psychic phenomena is that our brains can generate some mysterious rays that go out and change the world around us. But even if there were such rays, would they be so different from radio waves? Mind rays would, in the last analysis, still be susceptible to classical scientific analysis — and the feeling is that psychic phenomena, if they exist, should somehow lie outside normal science.

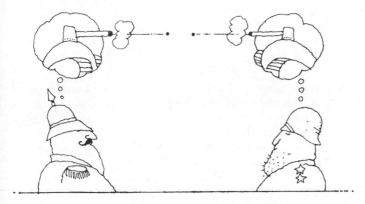

Fig. 188. The CIA has funded research for "psychotronic warfare."

No, I think that what people really expect from psychic phenomena is an ability to instantaneously affect external objects without any kind of cause-and-effect mediation at all. The wizard clenches his face, and on the other side of the galaxy a star goes supernova. Psychic phenomena are supposed to act faster than light, horizontally across spacetime.

PUZZLE 10.5

If time itself were bent into a vast circle, then one might hope to reach the past by traveling "around" time. But thinking about a universe in which time is a vast circle leads to some strange problems. Say, for instance, that you build a very durable radio beacon and set it afloat in space near the Earth. Is it possible that this beacon can last all the way around time? If it does, then once you set one afloat, how many more should you be able to detect? What if you decide to set your beacon afloat if and only if you detect no beacons out there before the launch?

Now, the new quantum experiments have established that — at least at the level of individual particles — there are significant and acausal cross-correlations between certain sequences of events. A typical experiment of this kind might consist of an energy source and two particle detectors at opposite ends of the laboratory. Each of the detectors prints out a random sequence of "yes or no" measurements. Although each sequence, taken by itself, seems random and meaningless, the experimenter finds that, taken together, the two sequences show a remarkable degree of similarity.

But this does not mean that one set of measurements is determining the other set. Two disparate regions of the universe may sometimes be in a kind of faster-than-light harmony, but it is meaningless to say that the events in one region are *causing* the events in the other region.

The great difference between synchronicity and telepathy is that we do not expect to be able to *control* synchronicity. The wizard's face clench and the star's explosion may happen at the same time — but it is foolish to say the star made the wizard grimace, as it would be to say that the wizard made the star explode. Synchronicity just happens.

Let's imagine a synchronicity at the human level. Say that Bob and Donna are lovers driven apart by the decrees of fate. They think of each other often, and still exchange letters for the first month or two. One week Bob happens to get out his old copy of Tom Wolfe's *Electric Kool-Aid Acid Test*. While reading it, he happens to hear a radio playing "Magic Man," by Heart. So far, so what. Bob is just randomly entertaining himself.

The next week Bob gets a letter from Donna: *"By the way, have you ever heard of 'Magic Man' by Heart? You fit the bill. This week I read a book called 'Electric Kool-Aid Acid*

PUZZLE 10.6

We discussed a number of paradoxes that arise from an ability to travel to the past. But just being able to communicate with the past leads to paradoxes as well. Suppose, for instance, that I have a magic telephone with the following properties: Whenever I pick up the receiver and dial "1," the magic telephone rings an hour earlier. I am thus able to telephone my past self. If I happen to hear the phone ringing, and pick up the receiver, I can expect to hear the voice of my future self. But now, what if at 9:00 I decide I will dial "1" at 11:00 unless I have received a call at 10:00?

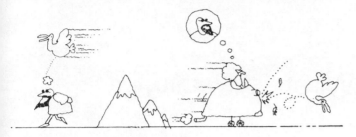

Fig. 189. Bob and Donna.

Test.' Do you know it? There's a girl in there called 'Mountain Girl.' Sometimes she and Kesey would talk for hours; he understood her better than anyone else; when I read that I thought of you & me."

What is poor Bob to think? Is Donna reading his mind? Are his actions controlling Donna's? No. Each of them is living out the random events of their lives. Yet these lives were once so close that there is still a degree of correlation between them.

This is a first-class universe we're in. It's loaded with symbolic events, deep meanings, and heavy coincidences. There are events that, for want of a better word, some people call telepathy. But there *is* a better word: the word is *synchronicity.* "Telepathy" suggests the idea of being able to exert control over coincidences. But surely the rough and tumble of life has taught all of us that any hope of full control is chimerical. Telepathy is a paranoid fantasy; synchronicity is a fact of life. As one 1960s slogan put it: "We don't have to get it together. It *is* together."

11

What Is Reality?

S TARTING with no preconceptions at all, what is the most reasonable model of the world that we can build up?

Only two things seem really certain: one exists, and one has perceptions. I may be a meat machine, a soul, an eye of God, a collection of ideas, or who knows what — but I am certain that I exist. I am the thing that writes these words. Of course, *you* may doubt whether I am real — perhaps you are only dreaming that you read this book — but you do know for certain that you yourself exist.

The fact that one has experiences is equally certain; to put it more neutrally, one cannot doubt that perceptions occur. In classical physics, one assumes that the perceptions are

Fig. 190. What is reality?

caused by objects in three-dimensional space; but, if you think about it, that is really quite an artificial assumption.

I do not perceive the world as stable objects, all equally present at all times. The world of immediate perceptions is all bits and pieces, rags and tags. The hiss of the radiator, a bad taste in my mouth, an ache in my hip, the red color of my typewriter, the clack of typing, the sky's pearl-gray glow, my glasses resting on my nose, rain dripping through the ceiling in the next room, birds in the sky, a car's tires on the street outside, the titles on my bookcase — this and that, now and again, a stew of perceptions, and stable reality is a mere construct! Concentrating now on my line of reasoning, I lose all awareness of the sky, rain, and cars; is it not reasonable to say that for this moment they cease to exist? Must an individual be like Atlas, who carries the whole world on his back?

Fig. 191. *Immediate perceptions.*

The Irish philosopher George Berkeley (1685–1753) advocated an idealistic philosophy called *immaterialism*. No one has written more eloquently on Berkeley than Borges does in his essay "A New Refutation of Time."

> Berkeley denied the existence of matter. This does not mean, one should note, that he denied the existence of colors, odors, tastes, sounds and tactile sensations; what he denied was that, aside from these perceptions, which make up the external world, there was anything invisible, intangible, called matter. He denied that there were pains that no one feels, colors that no one sees, forms that no one touches. He reasoned that to add matter to our perceptions is to add an inconceivable, superfluous world to the world. He believed in the world of appearances woven by our senses, but understood that the material world is an illusory duplication.

It is surprising to learn that such a seemingly perverse world view is embraced by modern physicists. In the words of John Wheeler, one of the grand old men of physics, "No elementary phenomenon is a phenomenon until it is an observed phenomenon." By this, Wheeler means that the rise of quantum mechanics has demolished the view that the universe sits "out there" while we sit back and observe it. The kinds of questions one asks — and the order one asks them in — has a profound influence on the answers one gets, and on the world view one builds up.

Fig. 192. *"No elementary phenomenon is a phenomenon until it is an observed phenomenon."*

What I would like to do now is build up a model of reality based on the notion that all that exists is the perceptions of

But you say, surely there is nothing easier than to imagine trees, for instance, in a park or books existing in a closet, and no body by to perceive them. I answer, you may so, there is no difficulty in it: but what is all this, I beseech you, more than framing in your mind certain ideas which you call *books* and *trees* and at the same time omitting to frame the idea of any one that may perceive them? But do you not you yourself perceive or think of them all the while? This therefore is nothing to the purpose: it only shows you have the power of imagining or forming ideas in your mind; but it doth not shew that you can conceive it possible, the objects of your thought may exist without the mind . . .

GEORGE BERKELEY,
A Treatise Concerning the Principles of Human Knowledge, 1710

Fig. 193. *Patterns of atoms.*

various observers. The "three space plus one time dimension" story is just one particular framework for organizing our sensations. We can — and do — equally well order our thoughts and impressions according to many other systems. Thoughts and memories having to do with "food" fall into one area, for instance, and these food thoughts are in turn organized according to various kinds of criteria. There are the good-food/bad-food axis, the sweet/salty axis, the raw/cooked, mine/not-mine, home-made/purchased, cold/hot, red/green, and endlessly many other axes.

What I am suggesting is that if we take sensations and thoughts as primary, then there is no reason to limit the "dimensions" of the world to the space and time dimensions involved in the motions of inanimate objects. Part and parcel of every object you see is what the object reminds you of, how you feel about it, what you know about its past, and so on. If we make an honest effort to describe the world as we actually live it, then the world grows endlessly more complicated than any simple 3-D picture. There is a feeling that the more we delve into the nature of reality, the more we will find. Far from being limited, the world is inexhaustibly rich.

We are used to taking left/right, forward/backward, up/down, past/future, and — perhaps — *ana/kata* to be the only possible dimensions. But why shouldn't cold/hot, nice/nasty, pretty/ugly, and all other distinctions have equal claim to being dimensions of reality? It is, after all, not enough to just tell a person where and when a given object is. One wants also to know if it is red, if it is good to eat, how much it costs, who else sees it, how much it weighs, and so on.

The traditional approach of science has been to explain sensory phenomena, such as color and warmth, in terms of the spacetime patterns of atoms. But who has ever really seen an atom? At best one might see some kind of electron-microscope photograph of a grainy dot. And one's *experience* of seeing such a photograph is in fact a collection of sensory phenomena, involving color, brightness, and so on. Matter may explain our perceptions, but it is our perceptions that tell us about matter.

I propose that we stop trying to explain our mental experiences in terms of invisibly tiny objects arranged in patterns in 3-D space. Instead let us take our actual thoughts and sensations as the truly fundamental entities. Think of a "dimension" as being any possible type of variation, category, or distinction. For each question you might ask about an object, there is a range of possible answers; we take each of these ranges to be an axis in the *true* "space" underlying our perceptions.

What shall we call this space? *Fact-space* sounds good. An entity of any kind is a glob in fact-space. To the extent that an entity's position on some axis is precisely given, the corresponding glob will have a narrow cross section in the direction of that axis. If an entity's properties are vague or indeterminate, then its glob will be hazy and spread-out. The world — the collection of all thoughts and objects — is a pattern spread out through fact-space.

Fig. 194. One-dimensional fact-space.

What is the pattern like? It will be helpful here to look at a lower-dimensional example. Imagine a Lineland inhabited by two Dots. Suppose for now that the Dots are at rest on their line of space, and suppose that they are so simple that all there is to know about any given Dot is its position. If we imagine looking at this Lineland from the outside, then it has a fact-space that is simply a line with two Dots on it. If for some reason the exact positions of the Dots are uncertain, then the globs representing their positions will be smeared out. In thinking of the smeared images, it is helpful to think of spots of light: take the smeared image to be a patch of light that is brighter at the center and shaded off at the edges.

There are, in practice, no really distinct boundaries between things. In terms of our model, this can be represented by supposing that no point between *A* and *B* is *totally* dark: the whole line is slightly lit, but as it has two particularly bright regions we say there are two distinct Dots.

If we suppose that the Dots have some properties other than position, then the fact-space takes on more dimensions. If we suppose, for instance, that the Dots can be at

You recall how it goes — one of the after-dinner party sent out of the living room, the others agreeing on a word, the one fated to be questioner returning and starting his questions. "Is it a living object?" "No." "Is it here on earth?" "Yes." So the questions go from respondent to respondent around the room until at length the word emerges: victory if in twenty tries or less; otherwise, defeat.

Then comes the moment when we are fourth to be sent from the room. We are locked out unbelievably long. On finally being readmitted, we find a smile on everyone's face, sign of a joke or a plot. We innocently start our questions. At first the answers come quickly. Then each question begins to take longer in the answering — strange when the answer itself is only a simple "Yes" or "No." At length, feeling hot on the trail, we ask, "Is the word 'cloud'?" "Yes," comes the reply, and everyone bursts out laughing. When we were out of the room, they explain, they had agreed not to agree in advance on any word at all. Each one around the circle could respond "yes" or "no" as he pleased to whatever question we put to him. But however he replied he had to have a word in mind compatible with his own reply — and with all the replies that went before . . .

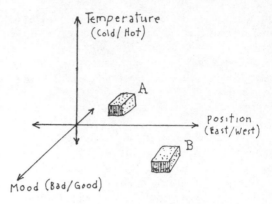

Fig. 195. Three-dimensional fact-space.

different temperatures and in different moods, then we get something like figure 195. Looking at it, we see that A is to the east of B, and that B, though colder than A, is in a better mood. Note also that the uncertainty about the Dots' exact mood is greater than the uncertainty about their temperature. But hold it; *whose* uncertainty am I talking about?

This is an important point. A fact-space like figure 195 is drawn as if some observer is standing outside the Dots' universe and measuring the Dots' properties. But this kind of attitude is not going to work if we want to talk about the actual world we live in. *We cannot observe our world from the outside.* If we take the "world" to be everything that exists, then there can be no outside observer!

Suppose we drop any notion of observing the Dots from the outside, and assume that the only real facts in Lineland involve *what the Dots know.* Assume again that the only property the Dots have is position. In this case the fact-space will have two axes: one axis for where A thinks the two Dots are, and one axis for where B thinks the two Dots are. If there were ten Dots, we would need ten axes, one for each Dot's opinion of the positions.

In figure 196 I have drawn the fact-space for two Dots, A and B, who notice their mutual positions. The lower patch represents A, and the other patch represents B. If we drop down vertically from the A-patch we get a narrowly defined location on the position-according-to-A axis. A knows quite well where A is. Notice, however, that A does not have a

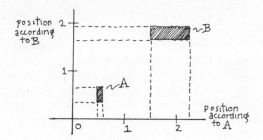

Fig. 196. Self-generated fact-space.

very good idea where *B* is. Studying the diagram further, we can see that *B* is a less "self-centered" Dot than *A* is. That is to say, if we run across horizontally from the *A*-patch and the *B*-patch, we find that *B* knows both Dots' positions with equal accuracy.

It is interesting to note that the quality of *A*'s "self-centeredness" is not something to be found just by looking at the *A*-patch. We notice this quality only by looking at the whole pattern in fact-space. What, by the way, shall we call a pattern in fact-space? Suppose we call it a *world state*, "state" being used here in the sense of "state of being" or "state of consciousness." Rephrasing the point just made, we might say that properties of an individual are bound up in the overall pattern of the world state.

If Dot *A* were to focus so entirely on its own location as to lose all sight of *B*, we would get a world state as shown in figure 197. For any individual at any given time, most of the world is going to be unknown or indeterminate, as *B* is relative to *A* here. Note, however, that the indeterminacy of *B*

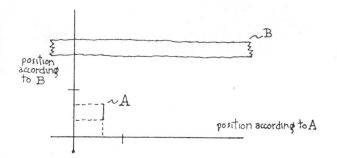

Fig. 197. A has no idea where B is.

In the real world of quantum physics, *no elementary phenomenon is a phenomenon until it is an observed phenomenon.* In the surprise version of the game no word is a word until that word is promoted to reality by the choice of questions asked and answers given. "Cloud" sitting there waiting to be found as we entered the room? Pure delusion!

JOHN A. WHEELER,
"Frontiers of Time," 1980

These ambiguities, redundancies, and deficiencies recall those attributed by Dr. Franz Kuhn to a certain Chinese encyclopedia entitled *Celestial Emporium of Benevolent Knowledge.* On those remote pages it is written that animals are divided into (a) those that belong to the emperor, (b) embalmed ones, (c) those that are trained, (d) suckling pigs, (e) mermaids, (f) fabulous ones, (g) stray dogs, (h) those that are included in this classification, (i) those that tremble as if they were mad, (j) innumerable ones, (k) those drawn with a very fine camel's hair brush, (l) others, (m) those that have just broken a flower vase, (n) those that resemble flies from a distance.

JORGE LUIS BORGES,
"The Analytical Language of John Wilkins," 1941

relative to A is still compatible with B's having a good knowledge of where B and A are.

It is worth noting here that this kind of model resolves one of the seeming paradoxes of immaterialism. Two people, let us say, are standing in a field. A rabbit runs past. One of the people sees the rabbit, the other doesn't. For the first person the rabbit has definite existence; for the second person the rabbit does not exist. How can one and the same rabbit both exist and not exist? In our fact-space model there is no problem. The glob representing the rabbit is shaped so as to have a sharp projection on one person's axis, and a fuzzed-out projection on the other person's axis.

Returning to our Lineland fact-space, note that if we were to add the properties of "temperature" and "mood" as before, then we would need a total of six axes: three axes for A's ideas about position, temperature, and mood; and three axes for B's opinions. In general, a world with P distinct properties and I distinct individuals will have P-times-I axes in its full fact-space. The state of the world can be thought of as a pattern of more or less distinct globs in a fact-space of very many dimensions.

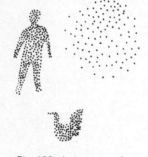

As the various individuals interact with each other, their knowledge about each other changes. The globs of light shift and merge, thicken and attenuate. We can think of the pattern as changing with the passage of time — alternatively, we can take each individual's time perception as an axis and have time change frozen right into the world state.

So far I have not discussed the issue of what kinds of things in our world are going to count as individuals. Just people? People and animals? People, animals, and plants? People, animals, plants, and robots? Should we include galaxies and rocks? Since we're going to get an unmanageably large collection anyway, my inclination is to be generous about this. I'm prepared to go ahead and let anything you

Fig. 198. A great number of individuals.

PUZZLE 11.1

Suppose that space and time are really just mental constructs. In general, the only reason we have for saying that one state of mind B comes after another state of mind A is that B includes a memory of state A, but A does not include a memory of state B. Under this definition of "before and after," would a person's perceptions necessarily fit into a linear time sequence?

like be an individual capable of certain kinds of "knowledge." A rock in a field doesn't know much, but it does know that there's something massive (the Earth) right under it. We can tell that it "knows" since if you lift it and let go, it falls right back down! Putting it a little more seriously, any given object, animate or not, "knows" or embodies information about a number of other objects.

This viewpoint — that everything is in some sense alive or conscious — has historically been known as hylozoism, or panpsychism. It is, of course, a viewpoint that is easy to ridicule. Can a garbage can feel love? Does a shoe think about mathematics? But the point is not that we are to think of objects as being "like people." The point is only that we might perhaps think of objects as integrative centers in fact-space, existing entities in the same sense that we are existing patterns ourselves.

So there are to be a great number of individuals in fact-space. And how many possible properties might there be?

Fig. 199. The world is like a game of "Infinity Questions" — there's no end to the number of questions you can ask.

There really seems to be no end to the number of properties individuals might be sensitive to. If you start asking questions about some individual object, it seems as though you might really keep on asking questions forever. So if fact-space is to have as many dimensions as the number of individuals *times* the number of properties, I think we may as well go ahead and say that fact-space is probably infinite-dimensional. Our world is a pattern in an infinite-dimensional, or "∞-D," space.

We are equal beings and the universe is our relations with each other. The universe is made of one kind of entity: each one is alive, each determines the course of his own existence.

The universe is made of one kind of whatever-it-is, which cannot be defined. For our purpose, it isn't necessary to try to define it. All we need to do is assume that there is only *one kind* of whatever-it-is, and see if it leads to a reasonable explanation for the world as we know it.

The basic function of each being is expanding and contracting. Expanded beings are permeative; contracted beings are dense and impermeative. Therefore each of us, alone or in combination, may appear as space, energy, or mass, depending on the ratio of expansion to contraction chosen, and what kind of vibrations each of us expresses by alternating expansion and contraction. Each being controls his own vibrations.

A completely expanded being is space . . . When a being is totally contracted, he is a mass particle, completely imploded . . . When a being is alternating expansion and contraction, he is energy . . . The universe is an infinite harmony of vibrating beings in an elaborate range of expansion-contraction ratios, frequency modulations, and so forth.

What we need to remember is that there is nobody here but us chickens. The entire universe is made up of beings just like ourselves.

THADDEUS GOLAS,
The Lazy Man's Guide to Enlightenment, 1972

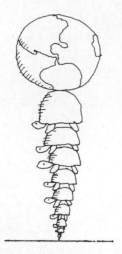

Fig. 200. Turtles all the way down.

The notion of ∞-D space is not something I just made up. The hyperspace philosophers of the late nineteenth century were well aware that the notion of higher dimensions leads to an infinite regress: Earth's 2-D surface is part of our 3-D universe. Our 3-D universe may well be the hypersurface of a 4-D hypersphere. Our 4-D hypersphere is a cross section of a 5-D curved spacetime pattern. Our 5-D curved spacetime is perhaps only a layer in a 6-D stack of alternate spacetimes. The 6-D stack may itself be warped and twisted in 7-D space; various versions of the stack may nestle together in 8-D space. Perhaps the whole 8-D space can be thought of as evolving along a nine-dimensional axis of supertime. And so on. Where can it stop? Only at infinity.

Hinton compared this to the kind of regress that arises if someone says the world rests on the back of a tortoise. What does the tortoise stand on? Another tortoise, which stands on another, which stands on another, ad infinitum. The theologian Arthur Willink found such an infinite regress of spaces exhilarating, and in his 1893 work *The World of the Unseen*, he argues that God lives in the ultimate ∞-D space:

> It is necessary to go farther, and to recognize a very wide extension of the idea of Higher Space, which is by no means exhausted when we have reached the conception of Space of Four Dimensions . . . When we have recognized the existence of Space of Four Dimensions there is no greater strain called for in the recognition of the existence of Space of Five Dimensions, and so on up to Space of an infinite number of Dimensions.
>   . . . And though it is impossible even to begin to imagine what

---

PUZZLE 11.2

The British writer J. W. Dunne felt that our dreams are made up of impressions taken equally from past and future. He claims in his 1927 An Experiment with Time *that the dreaming mind is able to rise up out of spacetime and see what lies ahead. This seems to lead to a sort of paradox: Suppose that I am to catch a plane Tuesday, and that, unknown to me, the plane is going to crash. Monday night, my dreaming mind sees into the future, and I have a horribly vivid dream of dying in a plane crash. Tuesday morning I am so badly shaken that I decide to postpone the trip. Tuesday evening, I watch the news and see that my intended plane did indeed crash, killing all aboard. The paradox is this: Since I did not in fact experience the plane crash, how could I have seen it as part of my future? Dunne's way out of this hinges on his claim that there is a second dimension of time. Can you fill in the details of his argument?*

the appearance of a material object in our Space may be to an observer in a much Higher Space, still it is evident that to him is presented a still more infinitely perfect view of its constituents than to an observer in any Lower region of Space. While to an eye in the Highest Space of all, an infinitely perfect revealing of the most hidden and secret things is of necessity presented.

This emphasizes very strongly what has been said about the Omniscience of God. For he, dwelling in the Highest Space of all, not only has this perfect view of all the constituents of our being, but also is most infinitely near to every point and particle of our whole constitution. So that in the most strictly physical sense it is true that in Him we live and move and have our being.

This is a very interesting passage, certainly one of the first philosophical uses of ∞-D space. A somewhat similar passage can be found in an 1886 essay, "On the Various Standpoints with Respect to Actual Infinity," by the German mathematician Georg Cantor:

The fear of infinity is a form of myopia that destroys the possibility of seeing the actual infinite, even though it in its highest form has created and sustains us, and in its secondary transfinite forms occurs all around us and even inhabits our minds.

It was Georg Cantor who first developed a mathematically rigorous treatment of infinity. Before Cantor, many mathematicians and philosophers had feared that infinity is basically a contradictory notion — but after Cantor, scientists were able to begin using infinities quite casually.

Early in the 1900s, mathematician David Hilbert drew on Cantor's work to develop a theory of infinite-dimensional spaces — spaces like the fact-space we have been talking about. Just as a point in a 3-D space can be mathematically represented by an ordered sequence of *three* coordinate numbers, a point in an ∞-D space is represented as an *infinite* sequence of numbers. There are various ways of going on to define things like angles and distances in these ∞-D

PUZZLE 11.3

According to quantum mechanics, if you don't keep an eye on a person, he or she soon becomes indeterminate for you: you no longer know what the other person is like. Yet if you ask the person questions, you will find him or her to have definite characteristics. Is there a contradiction here?

How can I express the final Teachings to which A Sphere led me? In my anguished fear of Death I begged him for some lasting Vision, some higher Truth to help me through the End.

SPHERE: It is difficult, oh Square. Before the Absolute we are both as Shadows. Space and Spacetime are but Fancies. Only in the Author is there final knowledge.

I: Where is the Author?

SPHERE: He is all around us, He is the mazy dark in which our patterns play. And even He is but a pattern in the unspeakable All.

I no longer knew whether I woke or slumbered. The Sphere's voice faded away, and all was confusion. I felt myself as but a Thought, a baseless fragment of some recurrent Dream. All around me I sensed my Dreamer's mind. Mustering my courage I cried out my plaint.

I: Can you hear me, my Lord?

DREAMER: And how! What time is it?

I: There is no Time — so says the Sphere.

DREAMER: Well, yeah. Not for you, anyway.

I: Return me to my fellows, oh my Author. Grant that the Hexagon forgives me.

DREAMER: I can do that. And thanks, I've enjoyed being with you. I hate to say good-bye.

I: But surely you will always be with me? Is not my World a fragment of your Mind?

spaces; the most commonly used type of mathematical ∞-D space is known as Hilbert space.

For a decade or so, the work on Hilbert spaces seemed to be simply another example of mathematicians pursuing abstraction for abstraction's sake. But in the 1920s, physicists Werner Heisenberg and Erwin Schrödinger discovered that the best way of interpreting quantum mechanics is to say that *particles are patterns in ∞-D Hilbert space.*

Since that time, mathematicians and physicists have developed an elaborate quantum-mechanical theory of the world as a pattern in an ∞-D Hilbert space. One of the great problems with this theory has been to find some actual significance for the ∞-D space involved. When classical physics presents us with equations involving four variables, we are able to understand these equations as being about the three space dimensions plus the one time dimension. But the mathematics of quantum mechanics seems somehow to have developed in the absence of any good underlying picture. There is no doubt that the quantum theory as it stands makes correct predictions about definite experiments. But no one seems to have a good feeling for what Hilbert space really means.

In describing our world of perceptions as a pattern in ∞-D fact-space, I have been trying to give the notion of ∞-D space some real content. The notion of fact-space is, in many ways, modeled on the quantum-mechanical Hilbert space. Given the protean nature of infinity, I would not be surprised to learn that the two spaces are really the same. It is important here to realize that the particular axes one lays down in a given space are quite arbitrary. Although a Californian and a New Yorker live in the same space, the directions they call "up" are slightly different. Space is given in itself, and without axes. Once we build up the world as a pattern in ∞-D fact-space, it could be that a simple redrawing of the axes would lead to the Hilbert space model. The important thing to remember is that the axes have no objective existence at all.

What is reality?

Take all your perceptions and all of mine, take everyone's thoughts and all the visions. In an infinite-dimensional space there is room to fit them all together; each is a piece of the infinite-dimensional One, and this One is reality.

Reality is indescribably rich and complex. Sometimes I forget this, and life goes gray. But the world is alive, and we are living parts of it. Thoughts are as real and important as objects. Every object is an endless source of wonder.

We don't know why we're here — we don't even know what we *are*. But we exist, and the world is going on. Our ordinary notions of space and time are just a convenient fiction. Higher dimensions are everywhere. There's no need to work for enlightenment; enlightenment is here and now, as close as the fourth dimension.

DREAMER: It's not *my* mind, really. I'm just filling in. Who knows who'll dream you next. You're the real immortal, Square, not me. You're an eternal Form.

For an instant I could see it All: the boundless Truth, the many Dreamers, and my own life's passionate play. And then I woke.

My father and A Hexagon were there with a policeman. The situation was dangerous, but so filled was I with Truth and Love that all was soon resolved. The four of us are fast friends to this day.

A SQUARE,
The Further Adventures of A Square, 1984

Puzzle Answers

ANSWER 1.1

If the two car shapes are of different *size*, then they can move past each other. The third dimension of distance-from-the-window is, in terms of images, represented by size. For the two-dimensional shapes moving in the window glass, size is a higher dimension. Imagine living in a three-dimensional analogue of such a world: you would be able to change your size at will, and you would be able to move "through" people of different sizes.

ANSWER 2.1

Self-gripping gut.

One way to keep poor A Square from going to pieces would be to have the type of gut depicted here. The projections on the upper half grip the knobs on the lower half, keeping A Square's body together. Food is passed down along the gut in the fashion of a barge moving through locks in a canal, with one after another of the barriers momentarily opening.

ANSWER 2.2

We would, in strict analogy to our world, expect two-dimensional creatures to crawl around on the rim of a disk: their planet.

ANSWER 3.1

If there were a 3-D space intersecting ours only in a plane, then we would see perhaps something like this: a plane of light slanting up from the ground into the sky, with odd-shaped globs drifting around in the plane, sliding up and down like holy Frisbees. The globs would be very thin and would feel solid to the touch.

ANSWER 3.2

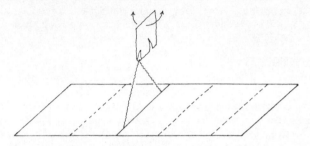

A Square scanning Flatland.

A Square's retina is a line segment designed to take in light from the plane of A Square's body. Now it seems that, looking down at Flatland from the third dimension, A Square would really see just those objects of Flatland that intersect the plane of his vision. The situation would be exactly like that of a Flatlander in a perpendicular world, as described in puzzle 3.1.

Now if A Square rocks back and forth, he can scan the various cross sections of Flatland and then mentally combine these sections to get a full 2-D image. By the same token, if you were in 4-D space looking at our world, you would see various planar cross sections of our world. With some effort, you would be able to combine these sections into a full 3-D image of everything, inside and out.

ANSWER 3.3

	Corners	Edges	Faces	Solids
Cube	8	12	6	1
Hypercube	16	32	24	8
Hyperhypercube	32	80	80	40

It is pretty easy to see that the number of corners will double each time we go up a dimension. But what about the other entries? How, without actually counting the lines in figure 34, do we know that the hypercube has 32 edges? The idea is that a hypercube is gotten by starting with a cube in initial position, moving the cube one unit *ana*, and then having the cube in final position. The initial and final cube each contribute 12 edges, and the cube's eight corners each trace out an edge during the *ana* motion. 12 + 8 + 12 = 32. Similar reasoning will justify the remaining entries.

ANSWER 3.4

The very bottom face must join up with the very top face, and the side faces on the bottom cube connect to the four open faces of the main part. This is easy to see if we consider the analogous process of folding a crucifix up into a cube.

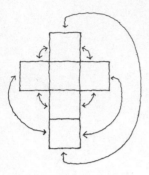

*Crucifix into cube.
(Answer 3.4)*

ANSWER 3.5

The formula is S^4, and the specific hypercube mentioned would have a hypervolume of 16 hypercubic feet.

The Greeks thought of numbers as specific geometrical magnitudes. For a given length S, S^2 stood for a certain square's area, and S^3 stood for the volume of a certain cube. Since they had no notion of a fourth dimension, they did almost no work with formulas or equations involving powers higher than 3. Only after the Renaissance did mathematicians feel confident enough of their algebra to work with higher-degree equations in a formal way.

ANSWER 3.6

If we go into 4-D space, then it is possible to find a fifth point (by moving *ana* from the tetrahedron's center), so that now all five points are the same distance from each other. These five points are the corners of a so-called pentahedroid.

In looking at this picture, we are to imagine that the central point is actually a bit further away in the fourth dimension, so that all the edges are really of equal length.

Just as a triangle is made up of three segments, and the tetrahedron is made of four triangles, the pentahedroid is made of five tetrahedra. Can you see all five?

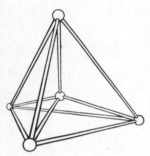

*The pentahedroid. (From
D. Hilbert and S. Cohn-
Vossen,* Geometry and
the Imagination.*)*

ANSWER 4.1

In cross section, a cube can appear to be a square, a triangle, a rectangle, or a hexagon. This is illustrated in a drawing by Claude Bragdon.

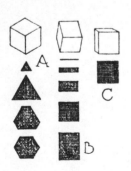

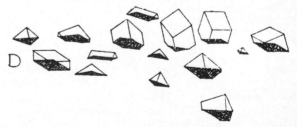

Cross sections of cubes. (From Claude Bragdon, A Primer of Higher Space.*)*

ANSWER 4.2

This illusion was sent to me by Orville L. Parrinello of Brazoria, Texas. At first it is a little hard to get, but the idea is that if you let the folded "staircase" reverse back and forth, the bug will alternate between being visible on the floor in front of Kilroy, or being hidden on the ceiling of the space behind Kilroy. The illusion is particularly interesting as it raises the idea that Kilroy's objective reality is not fixed at all. What if seemingly concrete facts about your life depended totally on how you look at things?

ANSWER 4.3

If A Square were transparent, then he could seem to reverse when seen edge-on, just as the transparent A Cube reverses in figure 51. The reversal would seem sort of as if the Square were "pulled through himself," just as a left glove can be turned into a right glove by turning it inside out.

Is this A Square or ǝɹɐnbS A? (Answer 4.3)

ANSWER 5.1

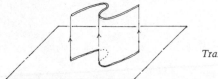

Trail of a line.

A line can't be knotted in 2-D space, because there's no way to have a line in Flatland cross above itself. And a line won't stay knotted in 4-D space because, as is discussed in chapter 5, the extra degree of freedom will cause any knot to slip through itself. Moving everything up a dimension, we expect that a plane can be knotted in 4-D space, but not in 3-D or in 5-D. How do you get a knotted plane? The idea is to start with a knotted line, and then imagine moving the knotted line *ana* out of space. The trail the line traces out will be a knotted plane. The plane, it is important to realize, is knotted, *but does not intersect itself.* Of course, if we move the knot in 3-D space, the trail does intersect itself, but since *ana* is perpendicular to every space direction, the 4-D trail will not come back on itself anywhere.

ANSWER 6.1

The turned-over Astrian would have his or her vibrator uselessly sticking up away from space. Such a person might seem to be without atmosphere or sensitivity, a zombie, a man without qualities, a burnt-out case. Is it interesting here to recall that some early thinkers, including René Descartes, thought that the pineal gland, located in the brain's center, might serve as a sort of third eye that perceives other people's auras or astral vibrations.

ANSWER 6.2

If an Astrian could dig his little 3-D thorn into underlying space hard enough, then he could oppose a force of gravity pulling him across space. In Charles H. Hinton's 1907 novel, *An Episode of Flatland*, this is exactly what the Astrians do:

> Just as the captain of a ship has an activity independent of the ship, so our souls have an activity independent of the body. Our souls can act on the alongside being [the aether] . . .
>
> If, filling my mind with devotion, I think of myself as soaring, as rising like an angel through the air, my soul does that which would make me rise, altering my direction by acting on the alongside being.
>
> If all men were to have the same thoughts, then all of them would tend to rise, and the united force would be very great, enough to influence the course of the earth in its orbit.

ANSWER 6.3

A Square executes a dangerous maneuver.

Cube snipped the piece of space containing A Square and turned the piece over. Now you might worry about the space of Flatland "popping" from having a hole in it. Well, perhaps A Cube took care of that by using something like embroidery rings to clamp down around the edges of the hole and of the cut-out piece. Or maybe there just was a hole in space for a while. What would it be like to have a hole in space? We'll get back to the question in chapter 7.

ANSWER 6.4

No, for we cannot tell whether the hole is moving through space. This is analogous to the fact that a particular matter bump does not serve to mark out some particular region of the fabric of space: as a wave moves across the water, its component water bits are constantly changing. We might compare a moving space hole to a bubble floating up through a liquid. Although the bubble's shape and size stay the same, the bits of liquid that lie at the bubble's boundary are changing as the bubble moves. An offbeat idea suggested by these considerations is that perhaps the smallest components of matter are not vortices or space bumps but actual *holes in space*.

ANSWER 6.5

The idea is that the galaxy's bump is on a line directly between us and the quasar. Light from the quasar to us can travel along two alternative shortest paths: one around either side of the galaxy's big space bump. This splitting of a quasar's image was definitely observed in 1979. (See Frederic Chaffee, "The Discovery of a Gravitational Lens," *Scientific American*, November 1980.) The phrase

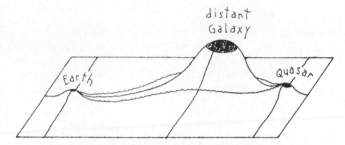

Three bumps in a row, and two shortest paths.

"gravitational lens" is an exciting way of expressing the fact that space curvature can bend light. It is amusing to think of a vast supertelescope based on gravitational lenses millions of miles across.

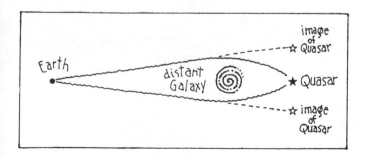

Top view of the two paths.

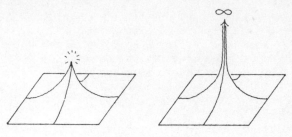

Two kinds of singularity.

ANSWER 6.6

A point-sized mass can be most simply represented as a sharp cusp in space. Alternatively, one might imagine this cusp to be pulled all the way out to infinity.

ANSWER 7.1

A geodesic on a sphere is a so-called great circle, that is, a circle, such as the equator, which is as big as possible. Relative to the sphere's surface, the equator is "straight" because it bends neither north nor south. A smaller circle, such as the Arctic Circle, can be seen to bend on the surface of the sphere and is not regarded as a geodesic.

ANSWER 7.2

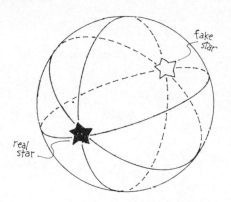

Virtual image of an antipodal star.

One natural explanation of a massless star would be to say that space is hyperspherical, and that the massless star is the virtual image of a real star at the opposite end of the universe. Unfortunately, even if space really is a hypersphere, we are not likely to actually observe any such "fake stars." The problem is that space is marred by medium-scale irregularities that will prevent the perfect focusing of the star's light rays at the point furthest away from the star. Another difficulty is that space contains clouds of dust here and there, dust which will absorb most of a star's light long before the light makes it halfway round the universe. Were it not for these two problems, we would in general expect to find a virtual image of any observed star at a diametrically opposite spot in the sky — provided, of course, that space really is hyperspherical.

ANSWER 7.3

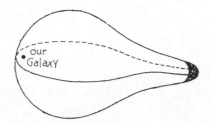

Finite universe with "central" mass.

It would look like a light bulb, or an ice cream cone: a ball with a big bulge on it. It is conceivable that our space actually has such a lopsided structure. If we were just about on the opposite side from the superstar that makes the bulge, then we wouldn't necessarily have to be able to see it directly. Space dust and intermediate space bumps could dissipate the monster's image. Such a model for our space is discussed in Paul Davies, *The Edge of Infinity*, 1983.

ANSWER 7.4

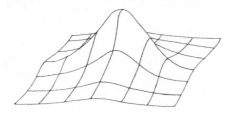

A curved space with natural distances.

We would get something like this: a square surface with a peak in the middle. The first and second pictures are two different ways of representing the same fact: there is more space toward the middle of this surface than one would ordinarily expect to find.

ANSWER 7.5

Square goes around the Möbius strip.

He turns into his own mirror image!

ANSWER 7.6

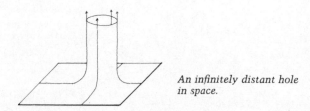

An infinitely distant hole in space.

The space near the hole is stretched up into an endless "chimney." No Flatlander ever gets to the end of the chimney; no Flatlander ever slides into the hole.

ANSWER 7.7

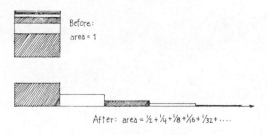

A finite area with infinite length.

The idea is to use an endless halving procedure to cut the square into infinitely many pieces: a piece of height $\frac{1}{2}$, one of height $\frac{1}{4}$, $\frac{1}{8}$, $\frac{1}{16}$, and so on. Zeno strikes again! If we line up the regions obtained, we get something that is infinitely long, but of unit area.

ANSWER 8.1

The two sheets of space might well snap apart. If the snap were too abrupt, it might do some damage to the Globber in question.

ANSWER 8.2

This is the type of connection represented in figure 112, although the holes in the spaces could be eliminated. The image is of a strip of space joining two distinct spaces together. The strip could, of course, be made very short by bulging the spaces down to meet each other at the surface of the mirror. (Keep in mind here that just as a mirror in our space is a piece of a plane, a mirror in Flatland is a piece of a line.) This kind of link between spaces is exactly what Lewis Carroll deals with in *Through the Looking-Glass*. Marcel Duchamp was also obsessed with the notion of mirrors as doors to alternative universes. He was struck by the fact that a point approaching a mirror has the choice, in principle, of either breaking through the mirror and continuing in normal space, or of moving out of our space and into the alternate space we see inside the mirror. Thus, for Duchamp, a mirror represented a sort of railroad switch where one chooses between two spaces: real space and mirror space. See Linda Dalrymple Henderson, *The Fourth Dimension and Non-Euclidean Geometry in Modern Art*, 1983.

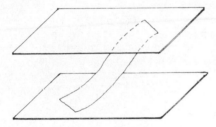

A link between two worlds.

A very vivid way to enhance the illusion that a mirror is a door into another world is to get a flashlight and go up to a mirror in a dark room. The laws of optics are such that if you shine the flashlight into the mirror, the image looks exactly as if the flashlight beam goes through the mirror and into a dark room on the other side of the mirror.

ANSWER 8.3

Install an Einstein-Rosen bridge in the basement, and have this hyperspace tunnel lead to an infinite space empty except for the endless plane of Flatland.

ANSWER 9.1

Take a small spherical balloon. Blow it up and then let the air out. The entire spacetime trail of the balloon's surface and inside is a solid hypersphere. The trail of the surface alone is the hypersurface of the hypersphere.

The speed with which the balloon should be inflated and deflated depends on what kind of conversion factor between space and time is adopted. In relativity theory the speed of light is used as a conversion factor; that is, a "meter of time" is taken to be the time it takes a light ray to travel one meter — about three ten-millionths of a second.

Constructing a hypercube in 4-D spacetime is even simpler. As Professor Tom Banchoff of Brown University puts it, "A hypercube in spacetime is just a *cube . . . for a while.*"

ANSWER 9.2

The "melting future" world view corresponds to the notion that future events exist, stored up and waiting for us. A uniform "now" moves forward with the passage of time, and instant after instant is permanently used up. In this viewpoint, past events are totally nonexistent. It is not uncommon for people to feel this way about their lives. Life here becomes a scarce resource that is consumed, and once something is over it doesn't matter at all. This is probably the least rewarding way possible to think about spacetime, as can be seen by thinking about the kind of personal philosophies inherent in the other three world views shown in figure 146. Whenever you cut yourself off from your past, you're in an extremely rootless and vulnerable position. But if you do throw out the past, you might as well throw out the future, too, and get totally into the "now."

ANSWER 9.3

Mr. Lee will put X halfway between A and C. The reason is that Lee will assume that it takes the light just as long to travel back from the other end of the platform as it took it to get there from his

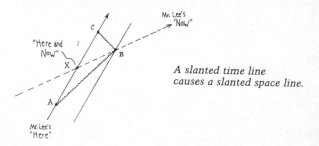

A slanted time line
causes a slanted space line.

end. It is natural for him to think this, in view of the assumptions (1) and (2) mentioned in the puzzle. We, of course, feel that it really takes the light longer to get from A to B than it takes it to get from B to C . . . But Mr. Lee will say we just think that because we're racing past him at half the speed of light!

If we draw a dotted line connecting B with the event X (which Mr. Lee says is simultaneous with B), then we get one of Lee's *lines of simultaneity*. This line corresponds to the space axis in a Minkowski diagram that Lee might draw; another way of putting it is that at the event X, this line will represent Lee's conception of "now" — just as his world line represents his concept of "here." It is possible to prove that in such a diagram, the observer's line of simultaneity will always tilt up by the same angle at which his world line tilts over.

ANSWER 9.4

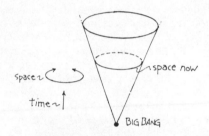

space z

time $\sim$

space now

BIG BANG

An expanding finite space.

The picture would be a sort of "conical" spacetime, as drawn here. The starting point is known as "the initial singularity," or as the Big Bang. Whether or not our space will eventually contract back to a point is unknown at present. Apparently it depends on how much mass is actually in our universe: if there is enough mass, then the gravitational forces will pull things back together.

ANSWER 10.1

To get into the sealed room, travel far, far into the future until you reach a time where the room's walls have crumbled. Step into the room's space and then travel back in time. Note that this is exactly analogous to moving *ana*, then moving across the location where the wall would be, and then coming back *kata* into the room's space.

Getting the food out of someone's stomach is a little trickier, since you yourself won't fit in the stomach. Let us imagine that the stomach in question belongs to your uncle Embry, snoozing off his

Christmas banquet on his bed upstairs. The thing to do is to go forward in time to later that afternoon when he's back out of bed. Take a little scoop and set it right on the bed where his stomach was. Now send the scoop back a few hours, to when he was there, and then bring it back to the empty bed. Do this a few times, and you can get all the chewed food out of Uncle's stomach. Hide it under his pillow so he has a nice surprise that night!

ANSWER 10.2

If we assume that the time machine goes straight backward in time, then "right here one week ago" is wherever the time machine appears when I send it one week back. One would not expect the Earth to still be "right here" a week ago, so traveling backward in time might well land one in empty space. Science fiction writers usually deal with this problem by somehow setting their time machine to track along the Earth's spacetime path.

It is interesting to realize that even without the paradoxes, time machines are already ruled out by relativity's basic assumption that there is no such thing as absolute rest or motion in space. Of course, if people somehow *could* build time machines, then we would probably just add some "except by using time machines" clause to the relativity principle. Or it might be that one would have to "aim" the time machine at some definite past object.

Note that just as a time machine can be used to define "right here," a machine that transmitted objects instantaneously could be used to define "right now." One could, for instance, spew out a whole lot of synchronized clocks all over space at the same instant. This also, of course, would violate relativity.

ANSWER 10.3

The point is that "instantaneous" is a relative concept. Relative to Earth, B happens at the same time as A, so one can travel instantaneously from A to B. Relative to the distant galaxy that moves away from us, C is simultaneous with B, so one can travel instantaneously from B to C. Combining the two trips takes one from A to C, and thus into one's own past.

ANSWER 10.4

Equip the rocket with a good robot brain so that after its hundreds-of-thousands-of-years-long journey, it can use the time machine to jump hundreds of thousands of years back in time and find the Earth. Once it finds the Earth again, it makes a small time jump to the right day (the day of launch) and lands.

ANSWER 10.5

The idea of a beacon that lasts all the way around time would lead to difficulties. If it drifted away from Earth never to return, then

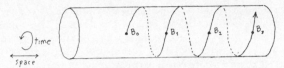

One *becomes* infinite?

there would be, it seems, endlessly many of them out there, as a consequence of the one launch! This seems nonsensical. The situation is particularly vicious if we suppose that B_1 sends out a signal that can inhibit the launch of B_0. A yes-and-no paradox!

More thought is required. Let us look at the individual particles that make up the beacon. If the universe really repeats itself, then it must be that each particle returns — at the end of each cycle — to its starting position. The world lines of particles are thus like rubber bands looped around the spacetime cylinder.

Now say that we are in such a circular-time universe, and say that we have built a durable radio beacon. The beacon has the form of a thick tangle of "bands" reaching around spacetime. Now, since we assembled the beacon — as opposed to just *finding* it — it must be that all of the beacon's component particles are going to end up back on Earth: as ore, as glass fragments, and so on. Therefore, we can logically conclude that in a circular-time universe, any ship we build and launch into space must eventually crash back onto Earth . . . so that its particles can be assembled back into the ship to be launched! It is, in other words, *impossible to build a truly indestructible object if time is circular!* For anything you build must in time disintegrate into pieces so that you can "again" build it.

ANSWER 10.6

We have a yes-and-no paradox here. I dial "1" at 11:00 if and only if I don't get a call at 10:00, but I get a call at 10:00 if and only if I dial "1" at 11:00. In other words, I get a call at 10:00 if and only if I

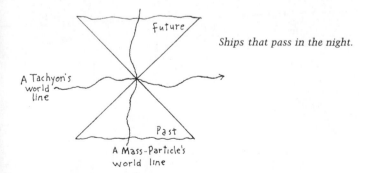

Ships that pass in the night.

don't get a call at 10:00. This particular paradox was first raised in a paper by G. Benford, D. Book, and W. Newcomb, "The Tachyonic Antitelephone" (1970). Gregory Benford is, by the way, a science fiction writer, as well as being a physicist. The "tachyons" his paper refers to are hypothetical particles that, unlike ordinary mass particles, *always* go faster than light. Benford's paper argues that since tachyons could be used to send messages into one's own past, it must be, even in principle, impossible to detect them. If, then, tachyons are real, they fill a sort of undetectable ghost universe whose time direction is in a sense perpendicular to our own time direction.

ANSWER 11.1

No. There are probably some sections of your life that do not involve any thoughts at all about the other sections. In "A New Refutation of Time," Borges argues that any state of mind that recurs in your life is *actually the same event.*

ANSWER 11.2

Dunne's idea is that at any instant we have a fixed future along a familiar time axis called T_1; but that as a second, higher sort of time, T_2, lapses, our future changes. Our actual time motion is a combination of T_1 motion, which we might think of as *moving into the future*, and T_2 motion, which we might call *moving into alternate worlds*. When I see something unpleasant ahead in this Monday's T_1 future, I am able to move in the T_2 direction into an alternate spacetime with a different T_1 future.

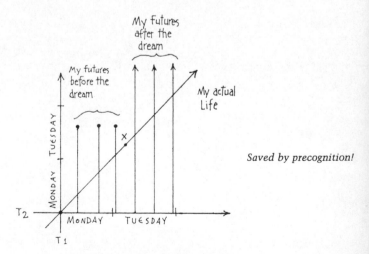

Saved by precognition!

Dunne tended to think in terms of infinite regresses. His notion of time leads to a regress, for if the dreaming mind can see ahead into T_1 future, it is not unreasonable to suppose that it can also see into T_2 future. And if you can see T_2 future, you can change it, which means that you are really moving along, let us say, T_3 time to a whole different T_1–T_2 plane! And, of course, there is no stopping at T_3: the regress goes on forever.

The reason Dunne was willing to embrace such a crazy system is that he was so dissatisfied with a simple Minkowski diagram's inability to depict our feeling that *time really passes*. As we discuss in chapter 11, any attempt to "animate" a Minkowski diagram by thinking of a "focus of consciousness" moving up along one's world line leads to the Dunne-style regress.

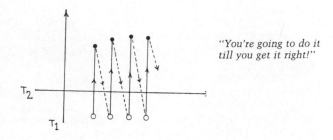

"You're going to do it till you get it right!"

In his 1885 story "An Unfinished Communication," Charles Howard Hinton describes a somewhat different sort of two-dimensional time. Hinton's concept is that one's life is repeated over and over, but with slight changes possible each time. After doing your life often enough, you finally get it right.

ANSWER 11.3

No, this is just another example of a region of fact-space having a vague position relative to one axis (your opinion), but a sharp position relative to another axis (the other person's opinion). If you are close enough to a person to be in some sense part of him (or her), then your opinion of him will evolve continuously with him. But if you are separated for a while and then brought back together, the other person's many possible states will seem to collapse down to one or two definite facts. In quantum mechanics this abrupt change is called "the collapse of the wave function." It is worth noting that, in certain relaxed states, you are yourself spread out in fact-space, even relative to your own axes. That is, if you are not presently asking yourself whether you are happy, then there is, at present, no definite answer to the question. Sharp or hazy, reality is just what it seems.

Bibliography

Abbott, Edwin Abbott. *Flatland: A Romance of Many Dimensions*. 1884. Reprint. New York: Barnes & Noble, 1983.

Borges, Jorge Luis. "A New Refutation of Time." In *Labyrinths: Selected Stories and Other Writings*. New York: New Directions, 1962.

——. *Borges: A Reader*. New York: E. P. Dutton, 1981.

Bork, Alfred. "The Fourth Dimension in Nineteenth-Century Physics." *Isis* 181 (1964).

Bragdon, Claude. *More Lives Than One.* 1938. Reprint. New York: Alfred Knopf, 1971.

——. *A Primer of Higher Space*. 1913. Reprint. Tucson: Omen Press, 1972.

Breuer, Miles J. "The Appendix and the Spectacles." 1928. Reprinted in *The Mathematical Magpie*, edited by Clifton Fadiman. New York: Simon & Schuster, 1962.

Burger, Dionys. *Sphereland*. New York: Apollo Editions, 1965.

Calder-Marshall, Arthur. *The Sage of Sex: A Life of Havelock Ellis*. New York: G. P. Putnam's Sons, 1959.

Campbell, Lewis, and William Garnett. *The Life of James Clerk Maxwell*. London: Macmillan, 1884.

Carroll, Lewis. *Through the Looking-Glass*. 1872. Reprint. New York: Random House, 1946.

Castaneda, Carlos. *A Separate Reality*. New York: Simon & Schuster, 1971.

Clifford, W. K. *Mathematical Papers*. London: Macmillan, 1882.

Conklin, Groff, ed. *Science Fiction Adventures in Dimension*. New York: Vanguard Press, 1953.

Davis, Andrew Jackson. *The Magic Staff*. New York: 1876.

Dewdney, A. K. *Two Dimensional Science and Technology*. Ontario: [No imprint,] 1980.

——. *The Planiverse*. New York: Poseidon Press, 1984.

Dolbear, A. E. *Matter, Ether and Motion.* Boston: Lee & Shepard, 1892.

Dunne, J. W. *An Experiment with Time.* 1927. Reprint. London: Faber & Faber, 1960.

Durrell, Fletcher. "The Fourth Dimension: An Efficiency Picture." In *Mathematical Adventures,* by Fletcher Durrell. Boston: Bruce Humphries, 1938.

Eddington, Arthur. *Space, Time and Gravitation.* 1920. Reprint. New York: Harper & Row, 1959.

Edwards, Paul. "Panpsychism." In *The Encyclopedia of Philosophy.* New York: Macmillan, 1967.

Einstein, Albert. "Ether and Relativity." In *Sidelights on Relativity,* by Albert Einstein. New York: E. P. Dutton, 1920.

Ernst, Bruno. *The Magic Mirror of M. C. Escher.* New York: Random House, 1976.

Faser, J. T., F. C. Haber, and G. H. Muller, eds. *The Study of Time.* Berlin: Springer-Verlag, 1972.

Gardner, Martin. *Relativity for the Million.* New York: Macmillan, 1962.

———. *The Ambidextrous Universe.* New York: Basic Books, 1964.

———. "The Church of the Fourth Dimension." In *The Unexpected Hanging,* by Martin Gardner. New York: Simon & Schuster, 1969.

———. "The Hypercube." In *Mathematical Carnival,* by Martin Gardner. New York: Alfred A. Knopf, 1975.

———. "Parapsychology and Quantum Mechanics." In *Science and the Paranormal,* by G. Abell and B. Singer. New York: Charles Scribner's Sons, 1981.

Gillespie, Daniel. *A Quantum Mechanics Primer.* New York: John Wiley, 1970.

Gödel, Kurt. "A Remark on the Relationship Between Relativity Theory and Idealistic Philosophy." In *Albert Einstein: Philosopher Scientist,* edited by Paul Schilpp. New York: Harper & Row, 1959.

Golas, Thaddeus. *The Lazy Man's Guide to Enlightenment.* Palo Alto: The Seed Center, 1972.

Gribben, John. *Timewarps.* New York: Dell, 1979.

Gustaffson, Lars. *The Death of a Beekeeper.* New York: New Directions, 1978.

Greenberg, Marvin. *Euclidean and Non-Euclidean Geometrics.* San Francisco: W. H. Freeman, 1974.

Hawking, S., and G. Ellis. *The Large Scale Structure of Space-Time.* Cambridge: Cambridge University Press, 1973.

Heinlein, Robert. "And He Built a Crooked House." 1940. Reprinted in *Fantasia Mathematica,* edited by Clifton Fadiman. New York: Simon & Schuster, 1958.

———. *Starman Jones*. 1953. Reprint. New York: Ballantine Books, 1978.

Henderson, Linda Dalrymple. *The Fourth Dimension and Non-Euclidean Geometry in Modern Art*. Princeton: Princeton University Press, 1983.

Hilbert, D., and S. Cohn-Vossen. *Geometry and the Imagination*. 1938. Reprint. New York: Chelsea, 1952.

Hinton, Charles Howard. *Selected Writings of C. H. Hinton*, edited by R. Rucker. New York: Dover, 1980.

Houdini, Harry. *A Magician Among the Spirits*. New York: Harper, 1924.

Huxley, Aldous. *The Perennial Philosophy*. New York: Harper & Row, 1944.

I Ching. Princeton: Princeton University Press Bollingen Series, 1950.

Jung, C. G. *Synchronicity*. Princeton: Princeton University Press Bollingen Series, 1973.

Kant, Immanuel. *Kant's Inaugural Dissertation and Early Writings on Space*. Chicago: Open Court, 1929.

Kaufmann, William J. *Relativity and Cosmology*. New York: Harper & Row, 1973.

Lewis, C. S. *The Lion, the Witch and the Wardrobe*. 1960. Reprint. New York: Collier Books, 1978.

Mach, Ernst. *The Science of Mechanics*. Chicago: Open Court, 1893.

Manen, Johan von. *Some Occult Experiences*. Chicago: Theosophical Publishing House, 1913.

Manning, Henry. *Geometry of Four Dimensions*. 1914. Reprint. New York: Dover, 1956.

Maxwell, James Clerk. *The Scientific Papers of James Clerk Maxwell*. 1980. Reprint. New York: Dover, 1963.

Minkowski, Hermann. "Space and Time." 1908. Reprinted in *The Principle of Relativity*, edited by A. Sommerfeld. New York: Dover, 1952.

Misner, C., K. Thorne, and J. Wheeler. *Gravitation*. San Francisco: W. H. Freeman, 1973.

Nabokov, Vladimir. *Look at the Harlequins*. New York: McGraw-Hill, 1974.

Neumann, John von. *Mathematical Foundations of Quantum Mechanics*. Princeton: Princeton University Press, 1955.

Nicholls, Peter, ed. *The Science Fiction Encyclopedia*. Garden City, N.Y.: Doubleday, 1979.

Ouspensky, P. D. *Tertium Organum*. 1912. Reprint. New York: Random House, 1970.

———. "The Fourth Dimension." In *A New Model of the Universe*. 1931. Reprint. New York: Random House, 1971.

Pagels, Heinz. *The Cosmic Code*. New York: Simon & Schuster, 1982.

Pearson, Karl. *The Grammar of Science*. London: Walter Scott, 1892.

Peebles, P. J. E. *Physical Cosmology*. Princeton: Princeton University Press, 1971.

Plato. "The Republic." In *The Dialogues of Plato*, translated by B. Jowett. New York: Random House, 1937.

Reichenbach, Hans. *The Philosophy of Space and Time*. 1927. Reprint. New York: Dover, 1958.

Rucker, Rudy. *Geometry, Relativity and the Fourth Dimension*. New York: Dover, 1977.

——. *Spacetime Donuts*. New York: Ace, 1981.

——. *Infinity and the Mind*. Boston: Birkhauser, 1982.

——. *The Fifty-Seventh Franz Kafka*. New York: Ace, 1983.

——. *The Sex Sphere*. New York: Ace, 1983.

Schoefield, A. T. *Another World; or, The Fourth Dimension*. London: Swann Sonnenschein, 1888.

Schubert, Hermann. "The Fourth Dimension." In *Mathematical Essays and Recreations*, by Hermann Schubert. Chicago: Open Court, 1898.

Stewart, Balfour, and Peter Guthrie Tait. *The Unseen Universe*. London: Macmillan, 1875.

Swenson, Loyd. *The Ethereal Aether*. Austin: University of Texas Press, 1972.

Taylor, Edwin, and John Wheeler. *Spacetime Physics*. San Francisco: W. H. Freeman, 1963.

Thorne, Kip. "The Search for Black Holes." In *Cosmology + 1*, edited by D. Gingerich. San Francisco: W. H. Freeman, 1977.

Wells, H. G. "The Time Machine." 1895. Reprinted in *Seven Science Fiction Novels of H. G. Wells*. New York: Dover, 1955.

Wheeler, John. "Frontiers of Time." In *Problems in the Foundations of Physics*, edited by N. di Franca and B. van Fraassen. Amsterdam: North-Holland, 1980.

Willink, Arthur. *The World of the Unseen; An Essay on the Relation of Higher Space to Things Eternal*. New York: Macmillan, 1893.

Wolf, Fred. *Taking the Quantum Leap*. San Francisco: Harper & Row, 1981.

Wolfe, Tom. *The Electric Kool-Aid Acid Test*. 1968. Reprint. New York: Bantam, 1969.

Zöllner, J. C. F. *Transcendental Physics*. Boston: Beacon of Light Publishing, 1901.

Index